Engineering and Design: An Introduction

WORKBOOK

Thomas Singer

PLTW Affiliate Professor
Sinclair Community College, OH

Patrick Lyons

Project Lead The Way Master Teacher
Fort Mill High School, SC

DELMAR
CENGAGE Learning™

Australia • Brazil • Japan • Korea • Mexico • Singapore • Spain • United Kingdom • United States

© 2010 Cengage Learning. All Rights Reserved. May not be scanned, copied or duplicated, or posted to a publicly accessible website, in whole or in part.

Engineering and Design: An Introduction: Workbook
Thomas Singer
Patrick Lyons

Vice President, Career and Professional Editorial: Dave Garza

Director of Learning Solutions: Sandy Clark

Senior Acquisitions Editor: James DeVoe

Managing Editor: Larry Main

Product Manager: Mary Clyne

Editorial Assistant: Cris Savino

Vice President, Career and Professional Marketing: Jennifer McAvey

Marketing Director: Deborah S. Yarnell

Marketing Manager: Jimmy Stephens

Marketing Coordinator: Mark Pierro

Production Director: Wendy Troeger

Production Manager: Mark Bernard

Content Project Manager: Mike Tubbert

Art Director: Bethany Casey

© 2010 Delmar, Cengage Learning

ALL RIGHTS RESERVED. No part of this work covered by the copyright herein may be reproduced, transmitted, stored, or used in any form or by any means graphic, electronic, or mechanical, including but not limited to photocopying, recording, scanning, digitizing, taping, Web distribution, information networks, or information storage and retrieval systems, except as permitted under Section 107 or 108 of the 1976 United States Copyright Act, without the prior written permission of the publisher.

For product information and technology assistance, contact us at
Professional & Career Group Customer Support, 1-800-648-7450

For permission to use material from this text or product,
submit all requests online at **cengage.com/permissions**
Further permissions questions can be emailed to
permissionrequest@cengage.com

Library of Congress Control Number: 2009925213

ISBN-13: 978-1-4180-6242-2

ISBN-10: 1-4180-6242-1

Delmar
5 Maxwell Drive
Clifton Park, NY 12065-2919
USA

Cengage Learning products are represented in Canada by Nelson Education, Ltd.

For your lifelong learning solutions, visit **delmar.cengage.com**

Visit our corporate website at **cengage.com**

Notice to the Reader
Publisher does not warrant or guarantee any of the products described herein or perform any independent analysis in connection with any of the product information contained herein. Publisher does not assume, and expressly disclaims, any obligation to obtain and include information other than that provided to it by the manufacturer. The reader is expressly warned to consider and adopt all safety precautions that might be indicated by the activities described herein and to avoid all potential hazards. By following the instructions contained herein, the reader willingly assumes all risks in connection with such instructions. The publisher makes no representations or warranties of any kind, including but not limited to, the warranties of fitness for particular purpose or merchantability, nor are any such representations implied with respect to the material set forth herein, and the publisher takes no responsibility with respect to such material. The publisher shall not be liable for any special, consequential, or exemplary damages resulting, in whole or part, from the readers' use of, or reliance upon, this material.

Printed in Canada
1 2 3 4 5 XX 10 09

© 2010 Cengage Learning. All Rights Reserved. May not be scanned, copied or duplicated, or posted to a publicly accessible website, in whole or in part.

Contents

© 2010 Cengage Learning. All Rights Reserved. May not be scanned, copied or duplicated, or posted to a publicly accessible website, in whole or in part.

© 2010 Cengage Learning. All Rights Reserved. May not be scanned, copied or duplicated, or posted to a publicly accessible website, in whole or in part.

© 2010 Cengage Learning. All Rights Reserved. May not be scanned, copied or duplicated, or posted to a publicly accessible website, in whole or in part.

© 2010 Cengage Learning. All Rights Reserved. May not be scanned, copied or duplicated, or posted to a publicly accessible website, in whole or in part.

© 2010 Cengage Learning. All Rights Reserved. May not be scanned, copied or duplicated, or posted to a publicly accessible website, in whole or in part.

© 2010 Cengage Learning. All Rights Reserved. May not be scanned, copied or duplicated, or posted to a publicly accessible website, in whole or in part.

Preface

This workbook was developed to support *Engineering Design: An Introduction* with real-world, hands-on activities that build basic skills for engineering design and provide opportunities to apply those skills in more challenging projects. The authors have combined their years of experience teaching Project Lead The Way's® Introduction to Engineering Design curriculum to produce a resource brimming with:

- Hands-on, directed design activities
- Drawing and sketching practice
- Math support
- Brainstorming and team development exercises
- Open-ended design problems and projects to provide greater challenges

Ample sheets of blank engineer's notebook paper, as well as orthographic and isometric grids are included for practice. As students complete the variety of activities in this workbook, they will be prompted to use these resources to develop their documentation skills.

Features of This Workbook

This text was developed to complement and support Project Lead The Way's® Introduction to Engineering Design curriculum, and can be used to support any project-based course in engineering design. The following features are built into each unit to help students apply the design process to achieve productive results.

BACKGROUND

Background sections help students develop and review the knowledge they need to perform the activities that follow.

TIP SHEETS

Tip sheets alert students to common pitfalls and provide helpful hints and motivating anecdotes to smooth students' journey toward successful design.

EXERCISES

At the core of this workbook are dozens of hands-on exercises that build essential design skills from math to brainstorming to sketching, drawing, and portfolio building.

© 2010 Cengage Learning. All Rights Reserved. May not be scanned, copied or duplicated, or posted to a publicly accessible website, in whole or in part.

Problem Sets

Units conclude with problem sets for additional practice at varying levels of rigor.

CASE STUDY

Golfsmith

This workbook includes a real-world case study in Unit 1 describing the design process used at Golfsmith to design and manufacture golf club heads. Thanks to the generous cooperation of design professionals at Golfsmith, the authors have been able to include rich information, technical drawings, and photos describing the Golfsmith design journey in detail. A series of real-world application activities involving golf club design and building continue throughout the manual, including prototyping, testing, and approval activities.

Additional Support for Teachers

Answers and solutions to the exercises in this text are provided for instructors on the Instructor Resources disk. The disk also includes a matrix correlating the activities in this workbook to the core text and to Project Lead The Way's® IED curriculum. To support the core text, the Instructor Resources disk also includes an Instructor's Guide, PowerPoint™ presentations, and a computerized text bank.

Engineering Design and Project Lead The Way, Inc.

This workbook is part of a series of learning solutions that resulted from a partnership forged between Delmar Cengage Learning and Project Lead The Way, Inc. in February 2006. As a non-profit foundation that develops curriculum for engineering, Project Lead The Way, Inc. provides students with the rigorous, relevant, reality-based knowledge they need to pursue engineering or engineering technology programs in college.

Project Lead The Way® curriculum developers strive to make math and science relevant for students by building hands-on, real-world projects in each course. To support Project Lead The Way's® curriculum goals, and to support all teachers who want to develop project/problem-based programs in engineering and engineering technology, Delmar Cengage Learning is developing a complete series of texts to complement all of Project Lead The Way's® nine courses:

Gateway to Technology

Introduction to Engineering Design

Principles of Engineering

Digital Electronics

Aerospace Engineering

Biotechnical Engineering

Civil Engineering and Architecture

Computer Integrated Manufacturing

Engineering Design and Development

To learn more about Project Lead The Way's® ongoing initiatives in middle school and high school, please visit www.pltw.org.

© 2010 Cengage Learning. All Rights Reserved. May not be scanned, copied or duplicated, or posted to a publicly accessible website, in whole or in part.

Acknowledgments

The authors and publisher wish to thank everyone who assisted in the development of the text, especially the reviewers and Master Teachers who provided valuable input. The authors owe special thanks to the team at Golfsmith, and especially to Jeff Smith, who shared his knowledge and examples to help make complex design points more understandable.

To my wife, Anita, and daughter, Rachel, for providing me the support and space to accomplish this endeavor, thank you

—Tom

To my wife, Teresa, and daughters, Hanna and Danielle, who inspire me every day, thank you

—Pat

© 2010 Cengage Learning. All Rights Reserved. May not be scanned, copied or duplicated, or posted to a publicly accessible website, in whole or in part.

UNIT 1
Engineering Design and Sketching

Skills List

After completing the activities in this unit, you should be able to:

- Identify the steps in the engineering design process

- Accomplish productive team-building activities

- Know the difference between patents, copyrights, and trademarks

- Complete a patent application

- Generate sketching projects in:
  o Orthographic projection
  o Isometric projection
  o Oblique (cavalier and cabinet) projection
  o One-point perspective
  o Two-point perspective
  o Three-point perspective

© 2010 Cengage Learning. All Rights Reserved. May not be scanned, copied or duplicated, or posted to a publicly accessible website, in whole or in part.

SECTION 1
The Design Process

The Design Process

The design process governs all aspects of how a product is developed and brought to market. As you begin the process of creating new ideas, you need to think about the materials that items are constructed from, and how each item will be disposed of or recycled at the end of its useful life. In other words, you need to think beyond the exciting birth of a product or tool to anticipate and document its complete life cycle. The design process helps organize your thoughts by providing placeholders.

Many variations of the design process are used in industry and described in textbooks today. A valid design process can have as few as 4 steps (Plan, Do, Study, Act), or more than 12 steps. The difference is the level of detail the design process uses to help guide you in the development process. Companies often adjust the process to suit their overall goals and timelines.

In this text you will study a 12-step design process, described by the National Society of Professional Engineers, as follows:

- Defining the problem
- Brainstorming
- Researching and generating ideas
- Identifying criteria and specifying constraints
- Exploring possibilities
- Selecting an approach
- Developing a design proposal
- Creating a model or prototype
- Testing and evaluating the design using specifications
- Refining the design
- Creating the design proposal
- Communicating processes and results

CASE STUDY

Golfsmith

Golfsmith is one of the major manufacturers of golf club equipment worldwide. They have generously allowed us to look inside their design process. In the sections that follow, you will see how Golfsmith takes new designs for golf club heads from concept into reality.

© 2010 Cengage Learning. All Rights Reserved. May not be scanned, copied or duplicated, or posted to a publicly accessible website, in whole or in part.

In any given year, Golfsmith (www.golfsmith.com) designs and manufactures up to 85 unique head designs. That's a lot by industry standards, since most other manufacturers of golf club heads produce 10 to 20 unique designs per year.

Each company uses its own design and manufacturing processes. Golfsmith's process is different from other golf club manufacturers. Design and manufacturing are not one-size-fits-all processes. Each company must look at its individual process.

The process for Golfsmith begins with the problem identification, which is the need to create a new unique golf club head design. As part of problem identification, criteria and constraints will be identified to guide the design development.

Golfsmith designers ask themselves a series of questions:

1. Who is the target golfer? Is the golfer highly skilled, average, or a beginner?

2. What is the performance objective? Does the golfer want to hit the ball the longest possible distance, control the ball, or achieve a combination of both?

3. Will the club head use traditional materials and shape (persimmon wood head design and constant-shaped irons) or an innovative design? See Figure 1-1.

FIGURE 1-1 *Golfsmith designs and manufactures a variety of different club shapes and types.*

4. What is the target cost of the club head? This question will determine the type of manufacturing process and the materials used in the golf club head construction.

5. What are the manufacturing process constraints? Will the club head be investment cast (created from a mold, lower cost) or forged (pressed into club form, higher cost) or a combination of casting and forging (cast in a mold and finished by pressing into the final shape)?

Through the brainstorming process sketches are created and then refined to just a few key ideas. See Figure 1-2.

FIGURE 1-2 *Design sketches for new golf club heads.*

© 2010 Cengage Learning. All Rights Reserved. May not be scanned, copied or duplicated, or posted to a publicly accessible website, in whole or in part.

Once the sketches are matched against the identifying criteria and specified constraints, the choices from the brainstorming process are refined down in the exploration of the possibilities, which ultimately lead to selecting an approach (or in this case a club head design).

Once the surviving designs are selected, a more accurate sketch is created so a prototype can be developed as part of the design proposal. See Figure 1-3.

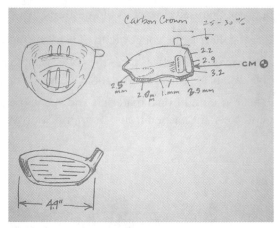

FIGURE 1-3 *Finished sketches of a new golf club head ready for prototyping.*

The prototype model is now created. Instead of creating the shape in a CAD system (this comes later in the process), the drawings are used to create accurate prototypes of the club head design in Golfsmith's own model design studio. See Figure 1-4. At the same time, the team develops a design proposal to obtain the necessary approvals to move from the design process to the manufacturing phase. When the prototype model is completed, it is again reviewed against the original design constraints originally developed to see if the criteria were met or exceeded.

FIGURE 1-4 *Prototyping a resin golf club driver using carving, sanding, and measurement tools.*

© 2010 Cengage Learning. All Rights Reserved. May not be scanned, copied or duplicated, or posted to a publicly accessible website, in whole or in part.

Once the prototype head is complete, a coat of paint is applied for the final evaluation of the design. See Figure 1-5. The evaluation group includes the design team, company management, and marketing as part of the design review before the final CAD drawings are developed and the manufacturing process begins.

FIGURE 1-5 *Painted resin prototype golf driver head.*

Once the initial prototype models are finished, they are sent to a machine shop and foundry in China, Taiwan, or Vietnam for laser scanning. See Figures 1-6 and 1-7. Laser scanning creates the surface mesh dataset that is transformed into a solid model. This solid model yields statistical data used in development, such as the center of gravity, centroid, and moments of inertia.

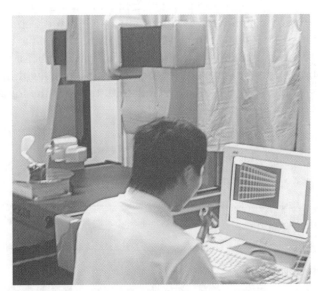

FIGURE 1-6 *Scanning a prototype golf club head to produce the 3-D CAD modeling data.*

© 2010 Cengage Learning. All Rights Reserved. May not be scanned, copied or duplicated, or posted to a publicly accessible website, in whole or in part.

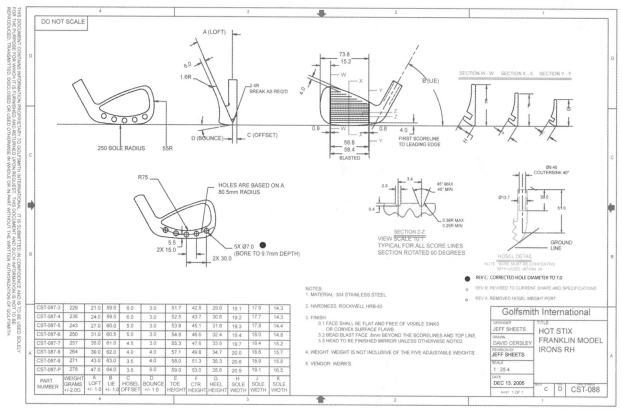

FIGURE 1-7 *CAD drawing of a golf iron head generated from the laser scanning process.*

The CAD models and the associated production drawings are sent back to Golfsmith for review. Once the models are approved, they are used to create casting models. The casting models are then used in the pre-production prototypes and the production molds.

Communication between the design team and the manufacturing companies is key at this stage. E-mail, instant messaging, online file sharing, and online Voice Over Internet Protocol (VOIP) tools like Skype are used to make sure the data and designs are being processed and manufactured to the standards set forth by Golfsmith.

Engineering samples (prototypes) are created in a limited production run to test against specific standards set internally by Golfsmith and externally by the United States Golf Association (USGA). This testing is part of the design reevaluation to see if refinements to the design are needed before the company releases it to production. The USGA requires all clubs to fall within a certain set of parameters before they can be sold. Golfsmith tests for the following design constraints:

- Materials tests
- Center of gravity location
- Durability of club head
- Sound at impact
- Cosmetics/graphics
- Centroid
- Moments of inertia

The pre-production prototypes are also tested on performance by both a robotic swinging arm (for consistency in use) and by Golfsmith employees on the

© 2010 Cengage Learning. All Rights Reserved. May not be scanned, copied or duplicated, or posted to a publicly accessible website, in whole or in part.

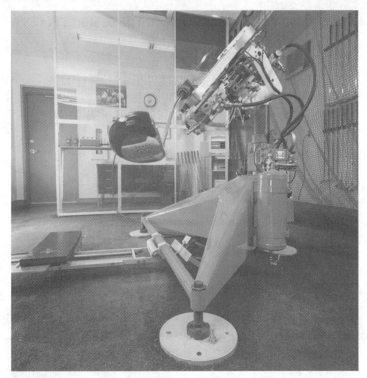

FIGURE 1-8 Robot testing a golf driver head and shaft combination for hitting accuracy.

driving range and golf course (for feel and weight as part of a finished club). See Figure 1-8. Once the testing is complete, communicating processes and results to the manufacturing group overseas completes the process with any final design adjustments.

This testing and revising process can take more than one cycle in most manufacturing scenarios. Golfsmith's process saves time by scanning a model to make the design. Less time is needed to revise club heads because the designer creates the prototype as the starting point instead of an ending point.

This design process works best for Golfsmith's design team. Other design and manufacturing organizations create the solid models through parametric modeling first and then develop the prototype from the CAD data. You'll find that different industries develop their own "styles" for design and manufacturing.

TIP SHEET

Don't Judge a Design Process

Do not judge a design/manufacturing process too swiftly. As an employee or consultant, you bring outside experiences into an organization. These experiences are an excellent resource for improvements. The key to success is to listen and watch to fully understand a process before interjecting ideas for changes or improvements. New employees often bring lots of ideas that initially get rejected—not because the idea wasn't good, but because it was too early in understanding the unique processes that occur at their new company.

© 2010 Cengage Learning. All Rights Reserved. May not be scanned, copied or duplicated, or posted to a publicly accessible website, in whole or in part.

EXERCISE 1.1 AN INTRODUCTION TO THE DESIGN PROCESS

Objective

To identify the steps in a design process and compare them to the 12-step process.

All design processes typically have a similar pattern; whether they are 4 steps, 7 steps, 10 steps, or 12 steps. The key is to be able to identify how these steps are used and grouped. See Figure 1-9.

- Defining the problem
- Brainstorming
- Researching and generating ideas
- Identifying criteria and specifying constraints
- Exploring possibilities
- Selecting an approach
- Developing a design proposal
- Creating a model or prototype
- Testing and evaluating the design using specifications
- Refining the design
- Creating the design proposal
- Communicating processes and results

Procedure

STEP 1 Do a search on the Internet for design processes. Your search will yield all types and styles of design process options. Select two different design process styles.

STEP 2 Group the 12-step design process steps to fit the new processes you have found through your research.

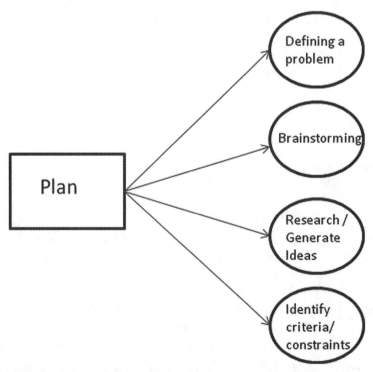

FIGURE 1-9 *The first four steps of the 12-step design process being mapped to the "Plan" step.*

© 2010 Cengage Learning. All Rights Reserved. May not be scanned, copied or duplicated, or posted to a publicly accessible website, in whole or in part.

EXERCISE 1.2 UNDERSTANDING THE DESIGN PROCESS

Objective

In this activity, you will study an integrated design process for developing renewable energy technologies and compare it to the 12-step process.

Procedure

STEP 1 Follow this link to access the Integration of Renewable Energy on Farms Web site: www.farm-energy.ca/IReF/index.php?page=design-process-idp

STEP 2 Study the Integrated Design Process (IDP) under the tab on the left side of the screen. Using our design process as a guide, develop a linking matrix between the two design processes.

Design Step	IDP Step(s)
1. Defining the problem	1. _____ 2. _____ 3. _____
2. Brainstorming	1. _____ 2. _____ 3. _____
3. Researching and generating ideas	1. _____ 2. _____ 3. _____
4. Identifying criteria and specifying constraints	1. _____ 2. _____ 3. _____
5. Exploring possibilities	1. _____ 2. _____ 3. _____
6. Selecting an approach	1. _____ 2. _____ 3. _____

continued

© 2010 Cengage Learning. All Rights Reserved. May not be scanned, copied or duplicated, or posted to a publicly accessible website, in whole or in part.

7. Developing a design proposal	1. _____ 2. _____ 3. _____
8. Creating a model or prototype	1. _____ 2. _____ 3. _____
9. Testing and evaluating the design using specifications	1. _____ 2. _____ 3. _____
10. Refining the design	1. _____ 2. _____ 3. _____
11. Creating the design process	1. _____ 2. _____ 3. _____
12. Communicating processes and results	1. _____ 2. _____ 3. _____

For additional information on the IDP, visit: www.iisbe.org/down/gbc2005/Other_presentations/ IDP_overview.pdf

© 2010 Cengage Learning. All Rights Reserved. May not be scanned, copied or duplicated, or posted to a publicly accessible website, in whole or in part.

SECTION 2
Team Building

BACKGROUND

Team Dynamics

Anytime you get two or more individuals together to work, they make a team. But with teaming comes good and bad. This is called *team dynamics*. The dynamics of a team changes over time, like any relationship, so the key is to manage it well. Everyone brings likes, dislikes, and biases into a team and the key is to use that variety to generate the best well-developed ideas. Individuals will have strong opinions and there may be clashes that are OK to a point.

TIP SHEET

Working Through Conflicts

Here are some general guidelines that will keep your team "one happy family."

- Make a list of everyone's skill set (best organized, good computer skills, most creative, etc.). This will help in determining the best tasks for each member (it is OK to create small work teams inside your larger team).

- Good teams start with communication via various methods. Not everyone is a talker; have team members submit ideas/thoughts on a variety of mediums: paper, via e-mail, video, in writing, etc.

- Listen first—don't open your mouth until you listen and think (good rule to help avoid conflicts).

- Conflicts are OK as long as it is not personal (never make a personal attack).

- Follow the design process; it will help guide the conversation.

- Set time limits for the design process steps.

- Record all ideas and options (either electronically or in writing).

TIP SHEET

Your Role in a Team

- Team Manager—oversees the progress, reports to the instructor.
- Team Recorder—records all the ideas and helps verify the design process is being followed.
- Team Specialist—has a specific task within a team, sometimes they work independently other time will be part of a work group within a team.

© 2010 Cengage Learning. All Rights Reserved. May not be scanned, copied or duplicated, or posted to a publicly accessible website, in whole or in part.

TIP SHEET

Five Steps in Team Evolution

Every team goes through the following five steps, but don't get stuck on any one step. Use the guidelines of good teaming to help you through the process.

1. Orientation (Forming)
2. Dissatisfaction (Storming)
3. Resolution (Norming)
4. Production (Performing)
5. Termination (Adjourning)

EXERCISE 1.3   WHAT YOU SEE IN AN IMAGE... MIGHT NOT BE THE TOTAL PICTURE

Objective

Your team will work together to record unique qualities of an object that other teams have not thought of. The winning team will have more unique items listed than any other team.

Materials

☐ A color picture of a carrot

☐ A stuffed carrot or a carrot model (optional)

☐ A bag of carrots (enough to share between the groups, small or regular size)

Procedure

Record the ideas that flow in each step.

STEP 1 ▸ Look at a picture of a carrot. Describe the carrot. Use all your senses and write down all aspects of the picture.

STEP 2 ▸ Look at a model of a carrot. Describe the carrot. Use all your senses and write down all aspects of the carrot.

STEP 3 ▸ Take a real carrot and do the same thing. List all characteristics that your team can find.

STEP 4 ▸ Now eliminate all the items that are the same on each team member's list. One team goes first and shares one of the unique items on its list. If one or more teams have that item on their list, scratch it off yours. Don't forget to snack on your carrots.

EXERCISE 1.4   WHAT'S IN A WORD?

Objective

Create a total of five unique sentences. The goal is to create wacky sentences.

Materials

☐ Engineer's notebook sheet

☐ Pencil

© 2010 Cengage Learning. All Rights Reserved. May not be scanned, copied or duplicated, or posted to a publicly accessible website, in whole or in part.

Procedure

STEP 1 Form teams of four or more members.

STEP 2 Select a topic from the sample list below, or supply your own topic.

STEP 3 One member takes notes for each sentence. Each team member in turn adds a word to build a sentence about the specific topic your team selected. The team goes around to each member until a sentence is complete. One complete sentence must be formed for each topic. None of the five sentences can start with the same word.

Sample Topics:

- An event happening at your school next week
- A current event happening in your city or state
- The coolest car you have seen
- The last funny movie you watched
- A sport that was on TV last night
- A favorite music artist or song
- Your cell phone manufacturer or service
- Your favorite TV show

© 2010 Cengage Learning. All Rights Reserved. May not be scanned, copied or duplicated, or posted to a publicly accessible website, in whole or in part.

SECTION 3
Documenting Your Work Through Trademarks, Patents, and Copyrights

BACKGROUND

How Do Engineers Protect Their Ideas?

Trademarks, patents, and copyrights are legal registrations designed to protect the rights of the designer (or original developer of the idea or work). Each protection is based on a different type of work, process, or idea that is developed. The protections differ between patents, copyrights, and trademarks under the law both in duration and coverage. It all starts with the use of an engineer's notebook. This legal document is typically the first way a unique idea first gains protection under the law. By following the appropriate steps in using an engineer's notebook, your rights are protected in the design development. One additional protection is if something does fail in a design you have a documented improvement process or information that may prove valuable. The goal is to present ideas so that someone else can read the sheets and understand the process and design ideas.

Make sure the following items are documented: Who, What, When, How, and Why.

WHO developed the idea?

- Everyone involved in the idea generation process should be included.

WHAT was the idea that was developed?

- Include all notes, sketches, ideas, contacts, Web sites used, and research information used.
- Any external documentation (like pictures) needs to be permanently mounted in the notebook.

WHEN was the idea first developed?

- All pages need to be dated.
- Make sure it is clear when the process began for the design.

HOW and WHY: This is the explanation of the idea, written in clear, legible language.

- Try to keep all your notes in one place, using sketch paper or scrap paper in the hopes that you will create a better drawing line your notebook, is not proper technique. Many times the ideas never make it into the notebook; you can always "line out" by drawing a single line (DO NOT scribble out)

© 2010 Cengage Learning. All Rights Reserved. May not be scanned, copied or duplicated, or posted to a publicly accessible website, in whole or in part.

and initial a sketch or passage. Using a single line will allow you to still read the section if the passage in error needs to be referenced or used later if there are additional changes.

TIP SHEET

Engineer's Notebook Reference

1. Keep several pages in the front of the notebook as a table of contents or quick reference for contact information (phone/e-mail). Note these pages as reference pages.

2. Any errors in the notebook can be corrected by documenting later a reference to the page and the change in the material to correct the error.

3. Make sure all the pages are initialed. Also, any changes or cross-outs need to be initialed each time a change is made.

4. Never leave any blank pages in the text. Use both sides of the paper.

5. Place an X in areas that are not being used on a page.

6. Always keep your notebook in a safe place. Do not let anyone else put anything in your notebook.

There are engineer's notebook pages found at the end of this workbook for you to document your ideas and sketches throughout the year. See Figures 1-10 and 1-11.

TIP SHEET

Engineer's Notebook

Never backdate or change any designs already in your engineering notebook. In industry an engineering notebook is a legal record of a design process. Designers and engineers typically use ink in notebooks so ideas cannot be changed.

TIP SHEET

Research

Remember: Wikipedia is not a source that you can quote; however, it is a source that can help you begin your search of a documented resource.

© 2010 Cengage Learning. All Rights Reserved. May not be scanned, copied or duplicated, or posted to a publicly accessible website, in whole or in part.

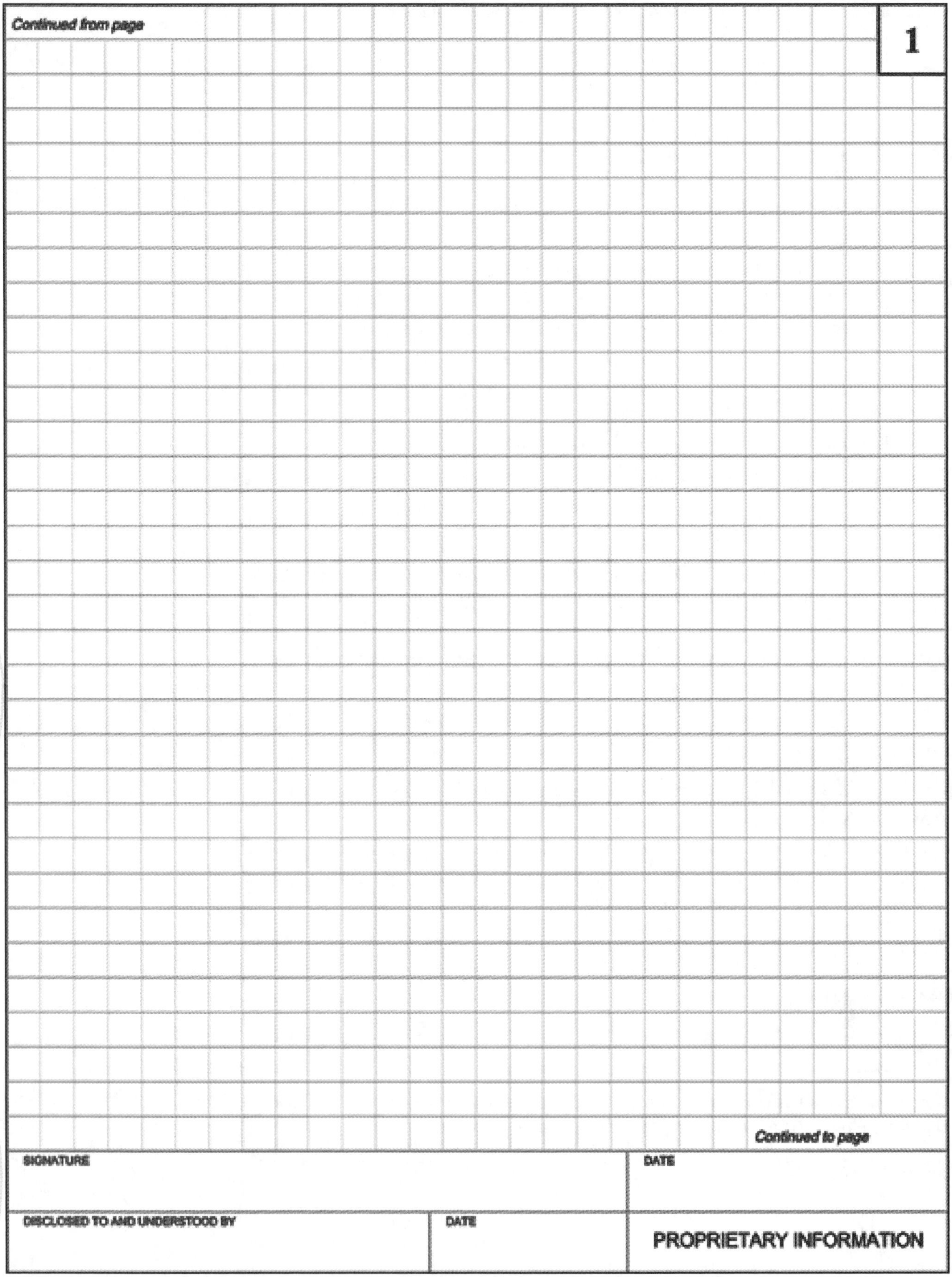

Continued from page ____

1

SIGNATURE

DATE

Continued to page ____

DISCLOSED TO AND UNDERSTOOD BY

DATE

PROPRIETARY INFORMATION

FIGURE 1-10 *Sample engineer's notebook page (more pages found in the back of the workbook).*

© 2010 Cengage Learning. All Rights Reserved. May not be scanned, copied or duplicated, or posted to a publicly accessible website, in whole or in part.

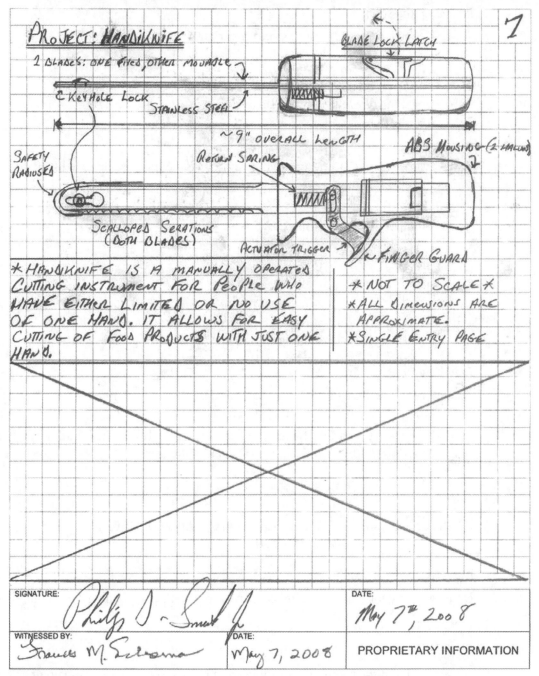

FIGURE 1-11 *Engineer's notebook sketched page of a design idea (sketch courtesy of Phillip Smith).*

EXERCISE 1.5   RESEARCHING TRADEMARK PATENT AND COPYRIGHTS

Objective

Research trademarks, patents, and copyrights and answer the following questions.

Procedure

STEP 1 ▶ Answer the following questions, individually, in complete sentences. Print two copies of your answers. Turn in one copy of your answers.

© 2010 Cengage Learning. All Rights Reserved. May not be scanned, copied or duplicated, or posted to a publicly accessible website, in whole or in part.

Individual Questions

- Define trademarks, patents, and copyrights.
- What does each cover?
- When do you need get one (explain for each type)?
- How long does each protection last?
- What are the submission requirements in the United States (or your country) for each?
- Find a patent that was created in your city or county. Provide the listing and any drawings.

STEP 2 ▶ Now break into three groups in your class. Each group will now take one of these three topics:

- Trademark
- Patent
- Copyright

STEP 3 ▶ Create an informational presentation on your group's specific topic that answers the individual questions and the following team questions:

Team Questions

- Are any of these protections recognized outside of the United States? If not, is there international protection of ideas?
- When was the earliest (patent, trademark, copyright) granted (and what was it for) in the United States? How about the world?

EXERCISE 1.6   FILLING OUT A PATENT AND COPYRIGHT APPLICATION

Objective

Understanding the paperwork involved in submitting a patent and a copyright.

Procedure

STEP 1 ▶ Visit the U.S. Patent Office online at www.uspto.gov/

STEP 2 ▶ Fill out the patent form, Application Data Sheet, at www.uspto.gov/ebc/portal/forms.htm, using your idea or design improvement.

STEP 3 ▶ Determine what the fees are for your patent application submission and write it on the top of the application sheet.

STEP 4 ▶ Drawings are covered by copyrights. Locate and fill out a copyright application, attach one of your CAD drawings.

STEP 5 ▶ Determine the fees for a copyright application, and write it on the top of the application sheet.

© 2010 Cengage Learning. All Rights Reserved. May not be scanned, copied or duplicated, or posted to a publicly accessible website, in whole or in part.

EXERCISE 1.7   WORKSHEET ON TRADEMARKS, PATENTS, AND COPYRIGHTS

Procedure

Research the following questions on the Internet and write complete answers in the space below:

1. What is the earliest and latest patent found on a spork design? Please provide details.

2. What company holds the patent for the pop-top openers found on cans?

3. What is the copyright information for this workbook?

4. Find and document a patent or copyright that covers an object you are specifically interested in this week.

5. What was the date that a crayon (coloring item) was first patented?

6. Name three patents Thomas Edison held.

7. Name the first African-American woman inventor to obtain a patent. (*Hint:* The patent was approved on July 14, 1885.)

© 2010 Cengage Learning. All Rights Reserved. May not be scanned, copied or duplicated, or posted to a publicly accessible website, in whole or in part.

SECTION 4
Introduction to Sketching

BACKGROUND

Why Sketch?

Have you ever had a great idea and then forgot what it was? If you get that idea on paper quickly, you will not only remember that idea, but you will be able to communicate that idea to other people. If you think about it, many things you own and use in your life began as a sketch. When an engineer or designer has an idea for something they want to create, they sketch that idea on paper first. Having the ability to get ideas on paper quickly and accurately is a skill that engineers and designers use all the time. In this unit we are going to learn about the importance of sketching and techniques you can use to get those great ideas on paper.

"I'm not an artist!" That is what most people say before they begin to learn how to sketch. As a small child you had to learn everything you know now, from reading and writing to math and science. Sketching is another learned skill like anything else. The more you sketch ideas on paper the better you will become. To create a great freehand sketch, all you need is a pencil and paper; no other tools are necessary.

TIP SHEET

Get This Straight

- A number-2 pencil will work fine for most sketches. These pencils work best if the lead is slightly dull and rounded.
- The amount of pressure you apply when sketching will determine the thickness of the line. If you push down hard, the line will be thick and heavy. If you apply a little pressure, the line will be lighter and easy to erase.
- When you first begin a sketch it is best to make light lines so that you can develop your idea on paper and then darken the lines when you are sure you have what you want.
- The best place to start when creating a sketch is how to draw a straight line.
- Remember, there is no need to use a straight edge such as a ruler.
- If you make a series of horizontal or vertical dashes or dots and connect them with a pencil, you have a straight line. What also helps when you are using this method is to look at the last dash, or the endpoint. Following these guidelines, you will have a straight line every time!

Alphabet of Lines

What are the different types of lines used in a sketch or drawing? See Figure 1-12.

© 2010 Cengage Learning. All Rights Reserved. May not be scanned, copied or duplicated, or posted to a publicly accessible website, in whole or in part.

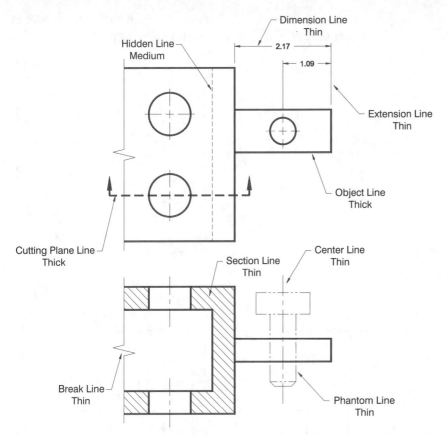

FIGURE 1-12 *The alphabet of lines is a variety of lines and what they represent.*

The specific line types tell a story in a drawing. Solid lines are surfaces that can be seen looking at that angle; hidden are surfaces that cannot be seen in that angle of view. Centerlines help locate features; phantom lines show alternate position of objects.

Break lines show where a drawing is removed due to size or shape; cutting plane lines indicate where an object is cut for sectioning.

Multiview (Orthographic) Sketching Principles

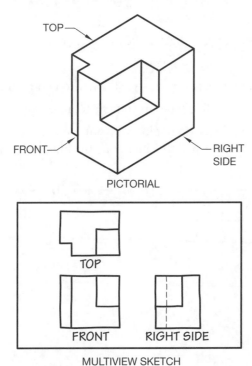

FIGURE 1-13 *Multiview drawing example.*

© 2010 Cengage Learning. All Rights Reserved. May not be scanned, copied or duplicated, or posted to a publicly accessible website, in whole or in part.

Multiview drawings need to have an "L"-shaped placement so that the reader of the drawing can interpret the surfaces. See Figures 1-13 through 1-17.

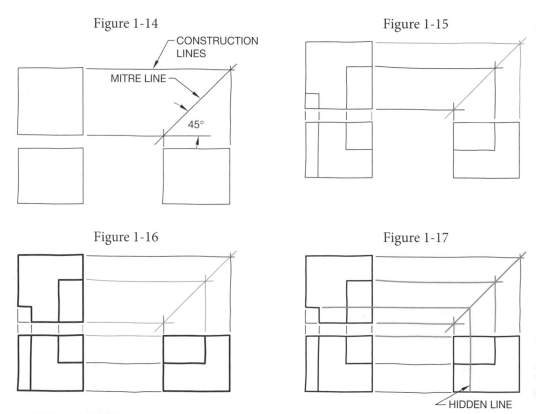

FIGURES 1-14 THROUGH 1-17   *Multiview sketching techniques.*

TIP SHEET

Dos and Don'ts

1. Don't feather lines use single-stoke line creation.
2. It's OK to have squiggles in your lines.
3. Surface alignment between views is the key to good multiview sketching.
4. Include the various line types in your sketching.
5. To draw straight lines, do not look at your hand, look at the point you need to move your pencil to. Then draw the line, and stare only at the destination point.

© 2010 Cengage Learning. All Rights Reserved. May not be scanned, copied or duplicated, or posted to a publicly accessible website, in whole or in part.

EXERCISE 1.8   HOW TO SKETCH A STRAIGHT LINE

Objective

To master the dash method and the dot method for creating straight lines.

Materials

☐ This workbook
☐ Pencil

Procedure

Follow the steps for the dash method

STEP 1 ▸ Make strokes right to left for a horizontal line. Use a continuous motion. Make a series of dashes and then connect them. See Figure 1-18.

STEP 2 ▸ Now you practice.

FIGURE 1-18   *One method of drawing straight lines begins with dashes and then filling them in.*

Procedure

Follow the step for the dot method

STEP 1 ▸ Place your pencil on the dot (on the right side).

STEP 2 ▸ Keeping your eye on the dot on the left side, move your pencil with one continuous stroke to make a straight line. See Figure 1-19.

STEP 3 ▸ Now you practice.

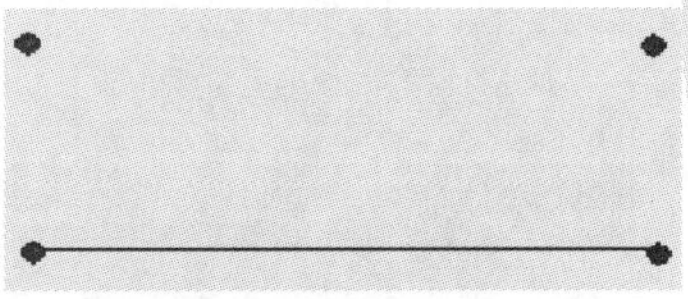

FIGURE 1-19   *Method two for drawing straight lines uses a "connect-the-dots" process.*

EXERCISE 1.9   HOW TO SKETCH A CIRCLE

Objective

Master the box method for drawing a circle.

Procedure

In addition to straight lines in a sketch, circles are also necessary and sometimes difficult to make. If you use a technique called the *box method*, circles and even ellipses will be easy to make. You begin by sketching a square of the circle's diameter with very light pencil lines; these types of lines are called *construction lines*. Construction lines are used to locate and develop a shape, and when the sketch is finished the construction lines can be erased.

© 2010 Cengage Learning. All Rights Reserved. May not be scanned, copied or duplicated, or posted to a publicly accessible website, in whole or in part.

STEP 1 Begin with a sketch of a square. See Figure 1-20.

FIGURE 1-20 *Draw a square.*

STEP 2 Make diagonal lines. See Figure 1-21.

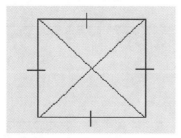

FIGURE 1-21 *Draw diagonal lines.*

STEP 3 Mark the midpoint of each side of the square. See Figure 1-22.

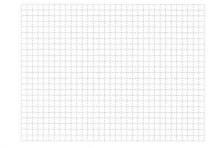

FIGURE 1-22 *Locate the midpoints on the sides of the square.*

STEP 4 Mark the radius of the circle on the diagonal lines. See Figure 1-23.

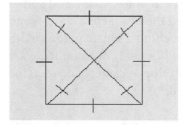

FIGURE 1-23 *Locate the radius on the diagonal lines.*

© 2010 Cengage Learning. All Rights Reserved. May not be scanned, copied or duplicated, or posted to a publicly accessible website, in whole or in part.

STEP 5 ▶ Draw the circle! See Figure 1-24.

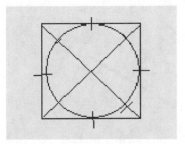

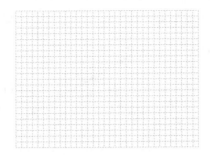

FIGURE 1-24 *Sketch the circle by connecting the line locations.*

STEP 6 ▶ When you are finished, erase the construction lines. You can use the same technique when sketching an ellipse by locating the major and minor axes. See Figure 1-25.

Draw your ellipse here.

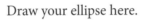

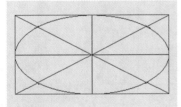

FIGURE 1-25 *Use a similar process of ellipse shapes.*

Pictorial Sketches

Now that we have some basic sketching techniques down it is time to begin to sketch some objects. An engineer or designer will create a sketch to capture an idea and then communicate that idea. This type of sketch is called a *pictorial sketch*. A pictorial sketch shows an object's height, width, and depth in a single view. We can classify pictorials has isometric, oblique, and perspective. It is important to note that a pictorial sketch is not a precise drawing like a multiview drawing. A pictorial drawing is like a picture; it just looks like the object. Later on we will look at creating multiview drawings using a technique called *orthographic projection*.

Isometric Sketches

Isometric means "equal measure." A pictorial sketch represents an object's height, depth, and width. In an isometric sketch you have three adjacent faces that share a single point. The edges that converge at this point will appear as 120-degree angles. The lines are equal, or 360 divided by 3, which is 120 degrees. If you draw a horizontal line and then a perpendicular line, the angle from the horizontal line is 30 degrees. See Figure 1-26.

© 2010 Cengage Learning. All Rights Reserved. May not be scanned, copied or duplicated, or posted to a publicly accessible website, in whole or in part.

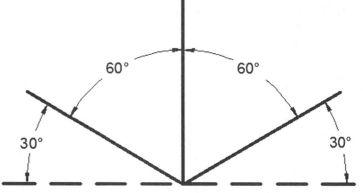

FIGURE 1-26 *Isometric drawing angles are 30 degrees and 90 degrees.*

Horizontal and 30-degree lines.

This is an isometric cube. Each face is separated by 120 degrees. See Figure 1-27.

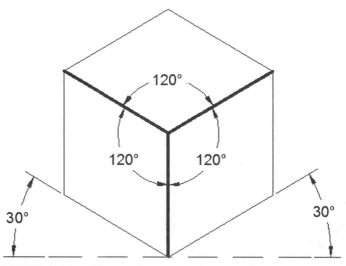

FIGURE 1-27 *This is an isometric cube. Each face is separated by 120 degrees.*

When creating a pictorial sketch, like an isometric sketch, it is a good idea to sketch a box with the overall dimensions. When you have the box made, sketch the features and complete the picture. See Figures 1-28 through 1-32.

Creating an isometric sketch.

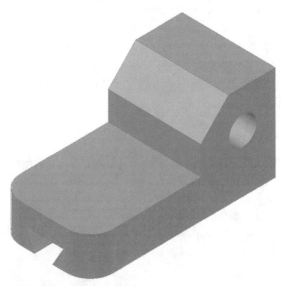

FIGURE 1-28 *Completed isometric shaded sketch.*

© 2010 Cengage Learning. All Rights Reserved. May not be scanned, copied or duplicated, or posted to a publicly accessible website, in whole or in part.

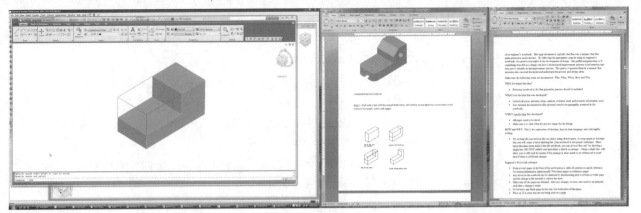

FIGURE 1-29 *Step 1: Use a very light touch and start with light lines (construction lines) to box in the length, width, and height.*

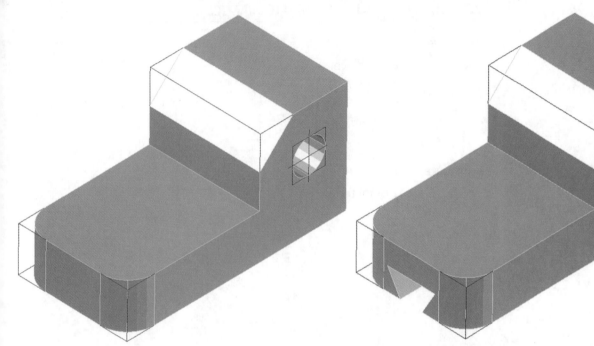

FIGURE 1-30 *Step 2: Rough in the features using box methods to lay out any circular features.*

FIGURE 1-31 *Step 3: Detail in the slot on the face of the object.*

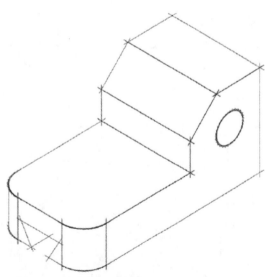

FIGURE 1-32 *Step 4: Darken all the object lines (no shading is required).*

Section 4: Introduction to Sketching   27

© 2010 Cengage Learning. All Rights Reserved. May not be scanned, copied or duplicated, or posted to a publicly accessible website, in whole or in part.

TIP SHEET

Isometric Tips for success

- Always start from a corner (either an external or internal corner that is formed on the sketch).

- Make sure the angled lines are parallel. Isometric sketches look incorrect when the angled lines are not consistent.

- Use very light lines so sections of the lines can be erased to give the illusion of being a solid "3D" style object.

EXERCISE 1.10   ISOMETRIC PRACTICE

Objective

Create an isometric sketch.

Materials

☐ This workbook

☐ Pencil

Procedure

Create the same isometric here.

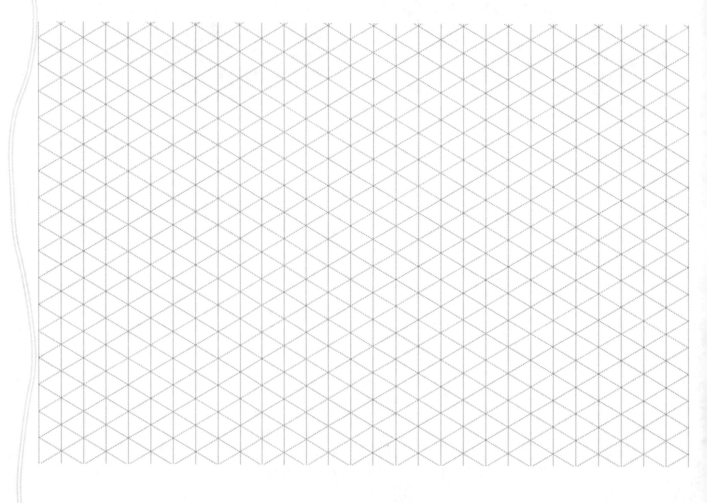

© 2010 Cengage Learning. All Rights Reserved. May not be scanned, copied or duplicated, or posted to a publicly accessible website, in whole or in part.

Oblique Sketches

An *oblique sketch* is defined as a sketch involving a combination of a flat front (true width and height) with depth lines receding at a selected angle, usually 30, 45, or 60 degrees. This type of pictorial drawing is used when you want to show one surface (or face) of an object without distortion. See Figure 1-33.

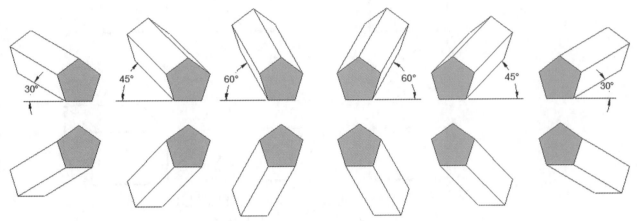

FIGURE 1-33 *Oblique angles come in all directions.*

To complicate things even further, oblique sketches can be broken down into oblique cabinet and oblique cavalier. In oblique cavalier sketches, all lines (including receding lines) are made to their true length. See Figure 1-34. In oblique cabinet, the receding lines are shortened by one-half their true length to compensate for distortion and to approximate more closely what the human eye would see. See Figure 1-35. For this reason, cabinet oblique drawings are the most common form of oblique drawings.

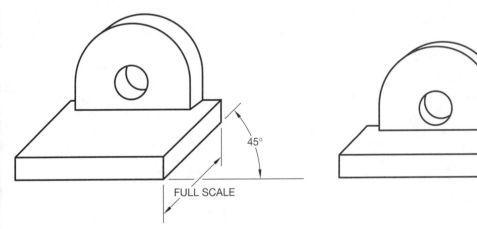

FIGURE 1-34 *Oblique cavalier.*

FIGURE 1-35 *Oblique cabinet.*

TIP SHEET

Oblique Sketches

To make an oblique sketch, start by drawing the front face, which is true width and depth. Next, use a 30-, 45-, or 60-degree line. See Figure 1-36.

© 2010 Cengage Learning. All Rights Reserved. May not be scanned, copied or duplicated, or posted to a publicly accessible website, in whole or in part.

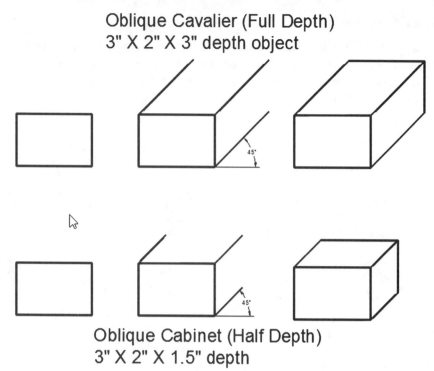

Oblique Cavalier (Full Depth)
3" X 2" X 3" depth object

45°

45°

Oblique Cabinet (Half Depth)
3" X 2" X 1.5" depth

FIGURE 1-36 *Examples of oblique styles.*

EXERCISE 1.11   OBLIQUE PRACTICE

Objective
Create an oblique sketch.

Materials
☐ This workbook
☐ Pencil

Procedure
Create the following oblique drawing on the grid provided. See Figure 1-37.

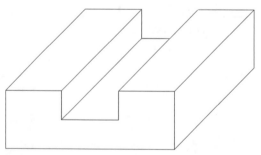

FIGURE 1-37 *Practice creating an oblique cavalier and oblique cabinet using the block image.*

© 2010 Cengage Learning. All Rights Reserved. May not be scanned, copied or duplicated, or posted to a publicly accessible website, in whole or in part.

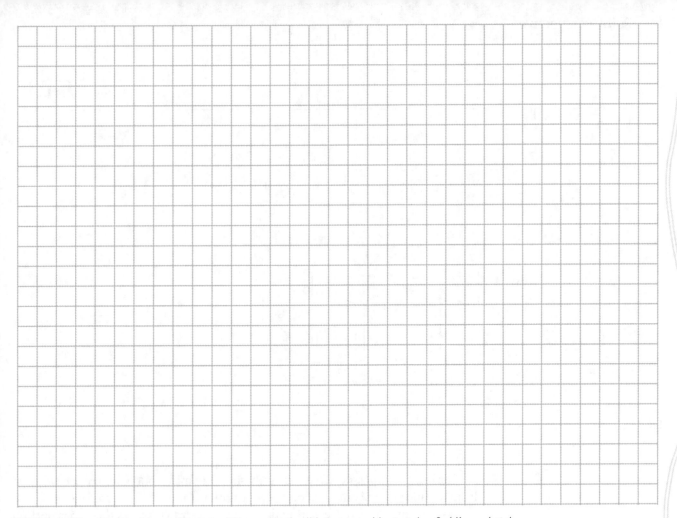

FIGURE 1-38 *(a) Create cavalier style oblique sketch. (b) Create cabinet style of oblique sketch.*

Perspective Sketches

The most realistic form of pictorial sketches is called *perspective sketches*. A perspective sketch uses vanishing points to provide the depth and distortion, that is, seen with the human eye. Perspective drawings can be drawn using one, two, and three vanishing points, also called one-, two-, and three-point perspective. Perspective sketches are in photographic proportion and easier for most people to visualize. See Figure 1-39.

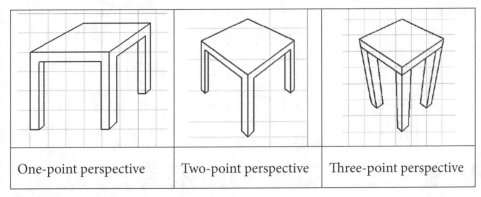

| One-point perspective | Two-point perspective | Three-point perspective |

FIGURE 1-39 *Various perspective styles.*

© 2010 Cengage Learning. All Rights Reserved. May not be scanned, copied or duplicated, or posted to a publicly accessible website, in whole or in part.

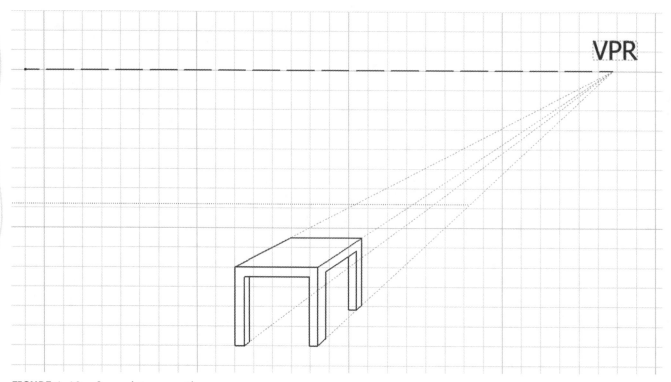

FIGURE 1-40 *One-point perspective.*

The one-point perspective is relatively simple to make, but is somewhat awkward in appearance when compared to other types of pictorials. See Figure 1-40.

A horizontal line, representing the horizon, is drawn across the upper portion of the paper. One vanishing point is identified somewhere on the horizon line. See Figure 1-41.

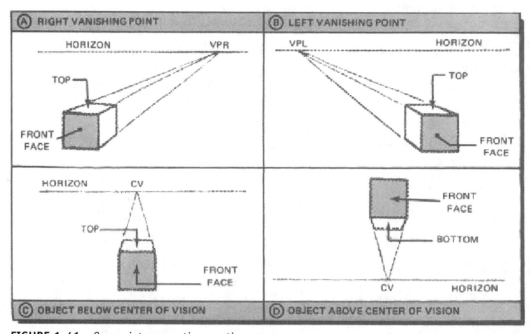

FIGURE 1-41 *One-point perspective creation process.*

© 2010 Cengage Learning. All Rights Reserved. May not be scanned, copied or duplicated, or posted to a publicly accessible website, in whole or in part.

EXERCISE 1.12   ONE-POINT PERSPECTIVE PRACTICE

Objective

Create a sketch in a one-point perspective.

Materials

☐ This workbook

☐ Pencil

Procedure

Use the grid to practice a one-point perspective sketch of a table. For an additional challenge, use the object shown here. See Figure 1-42.

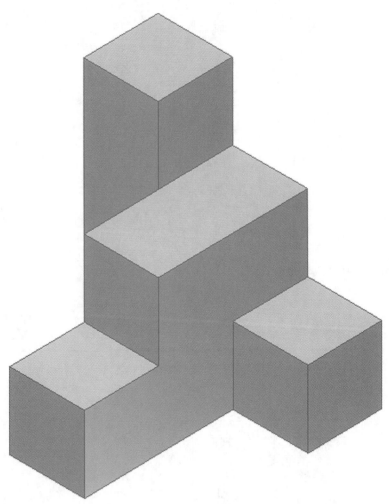

FIGURE 1-42   *Practice a one-point perspective sketch of a table.*

© 2010 Cengage Learning. All Rights Reserved. May not be scanned, copied or duplicated, or posted to a publicly accessible website, in whole or in part.

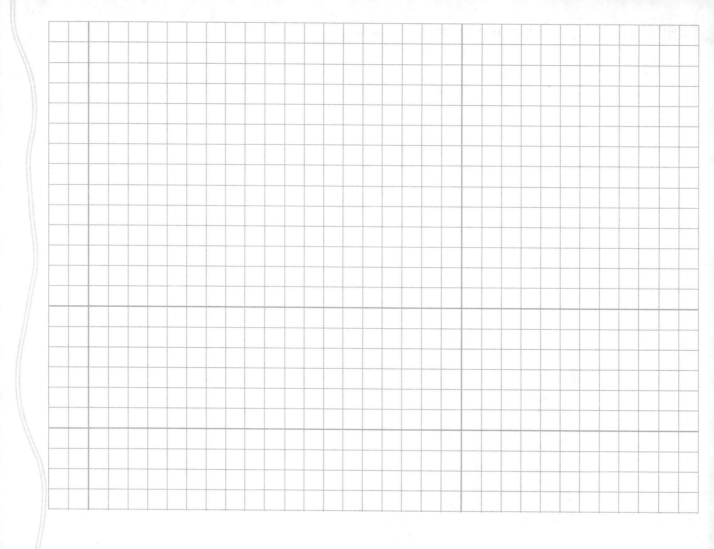

Two-Point Perspective

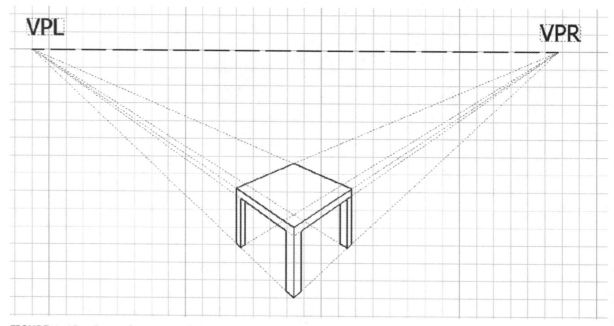

FIGURE 1-43 *Two-point perspective.*

© 2010 Cengage Learning. All Rights Reserved. May not be scanned, copied or duplicated, or posted to a publicly accessible website, in whole or in part.

Points and vertical construction lines are drawn to represent the overall width and depth of the object. Their locations must be estimated to make the overall dimensions of "the box" appear proportional. Once properly located, construction lines are drawn from the top points to the vanishing points on the horizon line. See Figures 1-43 and 1-44.

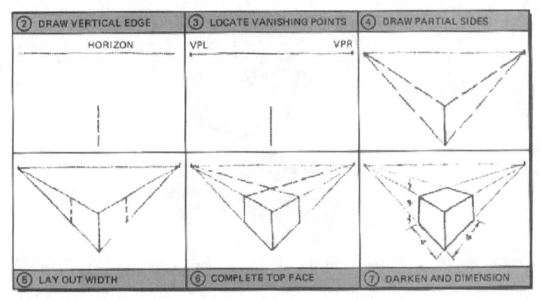

FIGURE 1-44 *Two-point perspective creation process.*

STEP 1 Sketch a light horizontal line for the horizon. Then position the object above or below this line so the right veritcal edge of the cube becomes the center of the sketch.

STEP 2 Draw a vertical line that represents the closest part of the cube.

STEP 3 Place two vanishing points on the horizon: one to the right and the other to the left of the object. Try to make them equal distance from the vertical line.

STEP 4 Draw straight lines from the corners of the vertical line to the vanishing points.

STEP 5 Lay out the width of the cube (true size along the angled lines), and draw vertical lines from those points.

STEP 6 Sketch the two remaining lines for the top, starting at the points where the two vertical lines intersect the top edge of the cube.

STEP 7 Darken all the object lines of the cube, and add dimensions if required.

EXERCISE 1.13 TWO-POINT PERSPECTIVE PRACTICE

Objective

Create a sketch in a two-point perspective.

Materials

☐ This workbook
☐ Pencil

© 2010 Cengage Learning. All Rights Reserved. May not be scanned, copied or duplicated, or posted to a publicly accessible website, in whole or in part.

Procedure

Use the grid to practice a two-point perspective sketch of a table. For an additional challenge, use the object shown here (Figure 1-45).

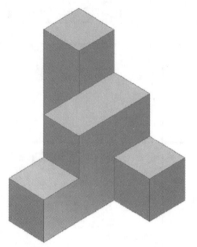

FIGURE 1-45 *Practice a two-point perspective sketch of the table.*

© 2010 Cengage Learning. All Rights Reserved. May not be scanned, copied or duplicated, or posted to a publicly accessible website, in whole or in part.

Three-Point Perspective

The three-point perspective gives the viewer either a worm's-eye, or bird's-eye view of an object. It is similar to two-point perspective, but now the vertical lines also vanish to a front vanishing point. See Figure 1-46.

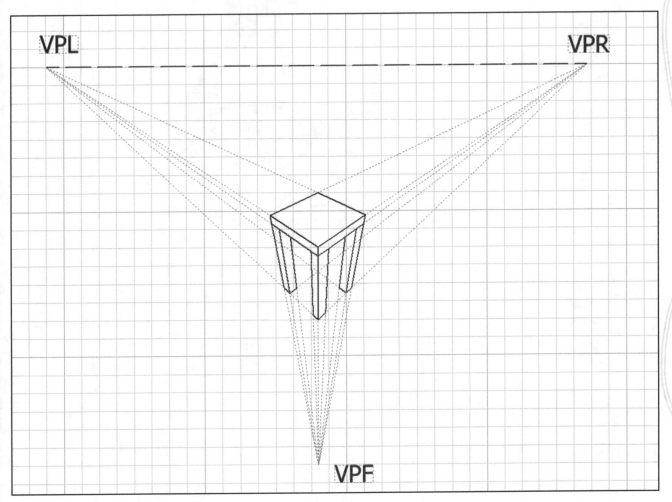

FIGURE 1-46 *Three-point perspective.*

EXERCISE 1.14 THREE-POINT PERSPECTIVE PRACTICE

Objective

Create a sketch in a three-point perspective.

Materials

☐ This workbook

☐ Pencil

© 2010 Cengage Learning. All Rights Reserved. May not be scanned, copied or duplicated, or posted to a publicly accessible website, in whole or in part.

Procedure

Use the grid to practice a three-point perspective sketch of a table. For an additional challenge, use the object shown here. See Figure 1-47.

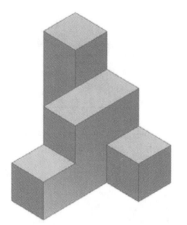

FIGURE 1-47 *Practice a three-point perspective sketch of a table example.*

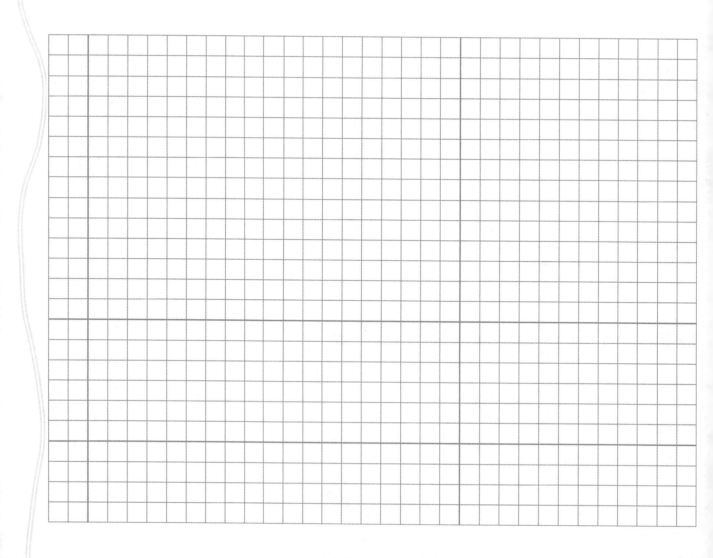

© 2010 Cengage Learning. All Rights Reserved. May not be scanned, copied or duplicated, or posted to a publicly accessible website, in whole or in part.

Practice Problems

Part A: Orthographic Sketching Problems

Insert the missing lines in the following orthographic views.

Problem 1-1

Create the orthographic sketched views using the appropriate line styles.

© 2010 Cengage Learning. All Rights Reserved. May not be scanned, copied or duplicated, or posted to a publicly accessible website, in whole or in part.

Problem 1-2

Create the orthographic sketched views using the appropriate line styles.

© 2010 Cengage Learning. All Rights Reserved. May not be scanned, copied or duplicated, or posted to a publicly accessible website, in whole or in part.

Problem 1-3

Create the orthographic sketched views using the appropriate line styles.

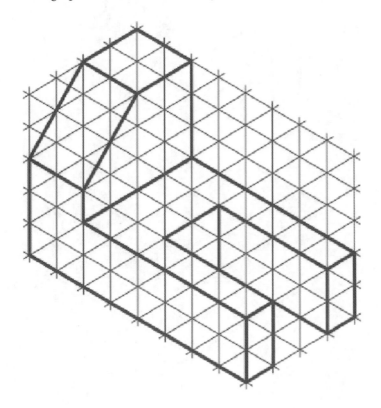

Problem 1-4

Create the orthographic sketched views using the appropriate line styles.

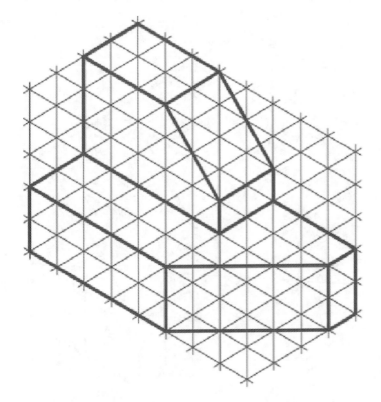

© 2010 Cengage Learning. All Rights Reserved. May not be scanned, copied or duplicated, or posted to a publicly accessible website, in whole or in part.

Problem 1-5

Create the orthographic sketched views using the appropriate line styles.

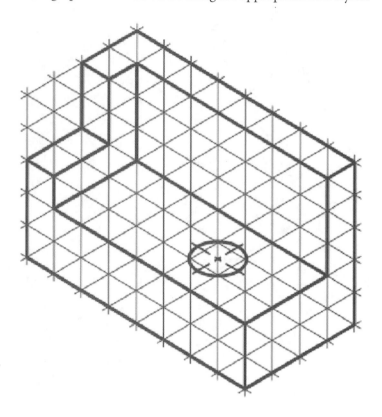

Problem 1-6

Create the orthographic sketched views using the appropriate line styles.

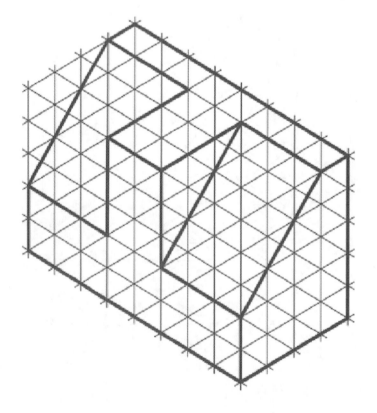

© 2010 Cengage Learning. All Rights Reserved. May not be scanned, copied or duplicated, or posted to a publicly accessible website, in whole or in part.

Problem 1-7

Create the orthographic sketched views using the appropriate line styles.

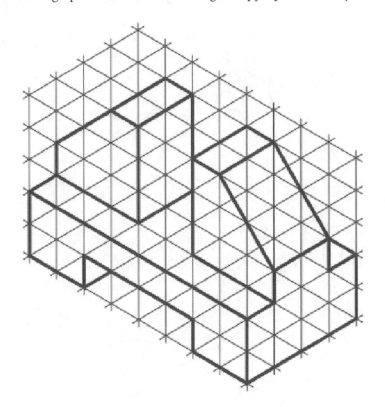

Problem 1-8

Create the orthographic sketched views using the appropriate line styles.

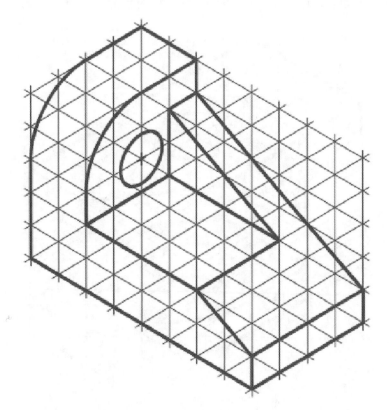

© 2010 Cengage Learning. All Rights Reserved. May not be scanned, copied or duplicated, or posted to a publicly accessible website, in whole or in part.

Problem 1-9

Create the orthographic sketched views using the appropriate line styles.

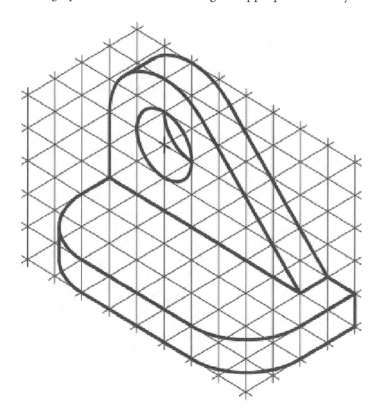

Problem 1-10

Create the orthographic sketched views using the appropriate line styles.

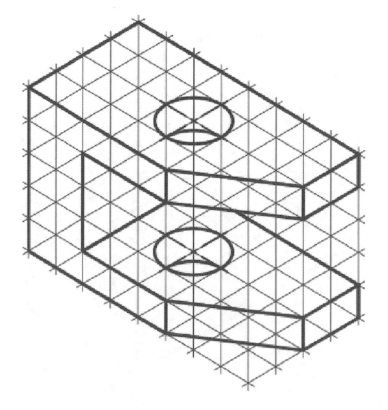

© 2010 Cengage Learning. All Rights Reserved. May not be scanned, copied or duplicated, or posted to a publicly accessible website, in whole or in part.

Problem 1-11

Create the orthographic sketched views using the appropriate line styles.

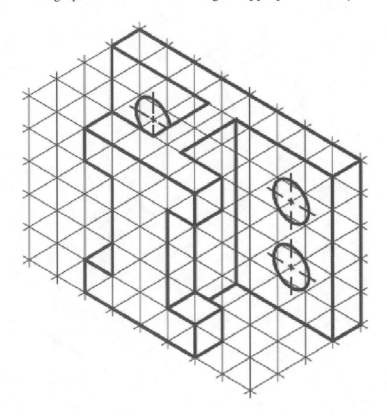

Problem 1-12

Create the orthographic sketched views using the appropriate line styles.

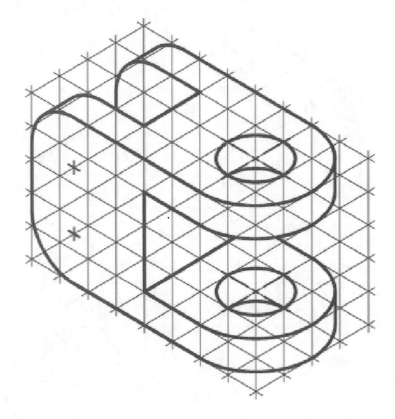

© 2010 Cengage Learning. All Rights Reserved. May not be scanned, copied or duplicated, or posted to a publicly accessible website, in whole or in part.

Problem 1-13

Create the orthographic sketched views using the appropriate line styles.

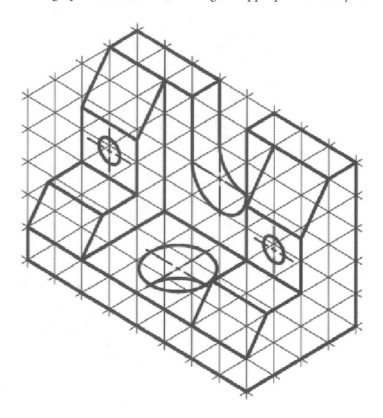

Problem 1-14

Create the orthographic sketched views using the appropriate line styles.

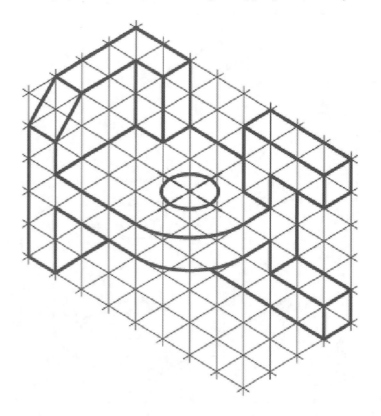

© 2010 Cengage Learning. All Rights Reserved. May not be scanned, copied or duplicated, or posted to a publicly accessible website, in whole or in part.

Part B: Isometric/Oblique and Perspective Sketching Problems

Create any of the following problems as a(n) isometric, oblique, or perspective sketch.

Problem 1-15

Create the isometric, oblique, or perspective views of the object as directed by your instructor.

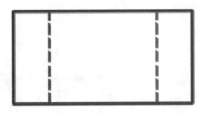

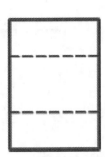

Problem 1-16

Create the isometric, oblique, or perspective views of the object as directed by your instructor.

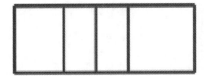

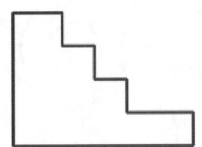

© 2010 Cengage Learning. All Rights Reserved. May not be scanned, copied or duplicated, or posted to a publicly accessible website, in whole or in part.

Problem 1-17

Create the isometric, oblique, or perspective views of the object as directed by your instructor.

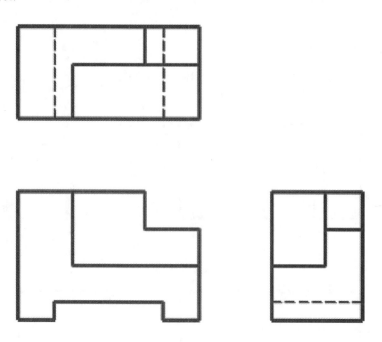

Problem 1-18

Create the isometric, oblique, or perspective views of the object as directed by your instructor.

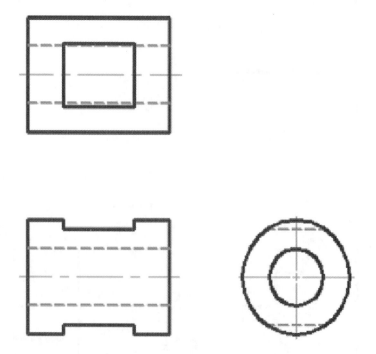

© 2010 Cengage Learning. All Rights Reserved. May not be scanned, copied or duplicated, or posted to a publicly accessible website, in whole or in part.

Problem 1-19

Create the isometric, oblique, or perspective views of the object as directed by your instructor.

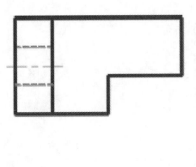

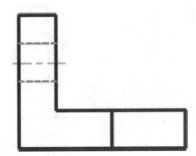

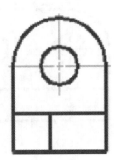

Problem 1-20

Create the isometric, oblique, or perspective views of the object as directed by your instructor.

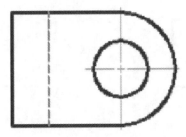

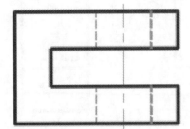

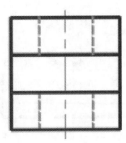

© 2010 Cengage Learning. All Rights Reserved. May not be scanned, copied or duplicated, or posted to a publicly accessible website, in whole or in part.

Problem 1-21

Create the isometric, oblique, or perspective views of the object as directed by your instructor.

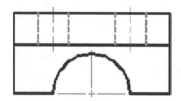

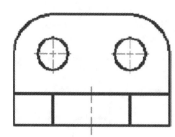
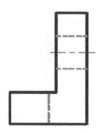

Problem 1-22

Create the isometric, oblique, or perspective views of the object as directed by your instructor.

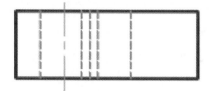

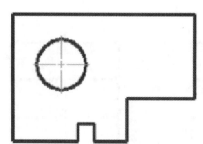
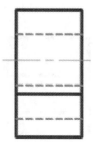

© 2010 Cengage Learning. All Rights Reserved. May not be scanned, copied or duplicated, or posted to a publicly accessible website, in whole or in part.

Problem 1-23

Create the isometric, oblique, or perspective views of the object as directed by your instructor.

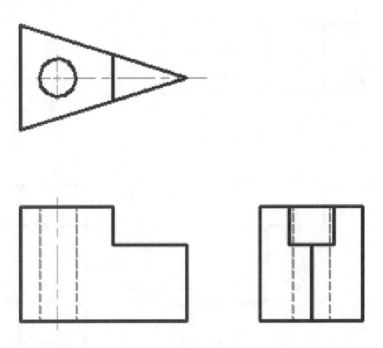

Problem 1-24

Create the isometric, oblique, or perspective views of the object as directed by your instructor.

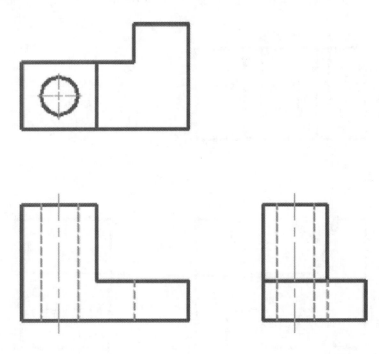

© 2010 Cengage Learning. All Rights Reserved. May not be scanned, copied or duplicated, or posted to a publicly accessible website, in whole or in part.

Problem 1-25

Create the isometric, oblique, or perspective views of the object as directed by your instructor.

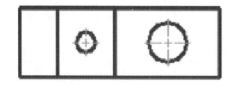

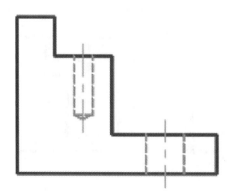

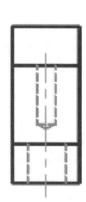

Problem 1-26

Create the isometric, oblique, or perspective views of the object as directed by your instructor.

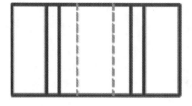

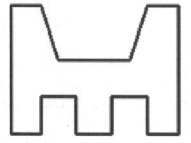

© 2010 Cengage Learning. All Rights Reserved. May not be scanned, copied or duplicated, or posted to a publicly accessible website, in whole or in part.

Problem 1-27

Create the isometric, oblique, or perspective views of the object as directed by your instructor.

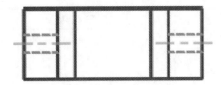

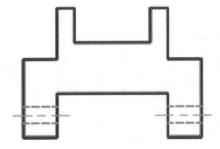 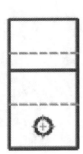

Problem 1-28

Create the isometric, oblique, or perspective views of the object as directed by your instructor.

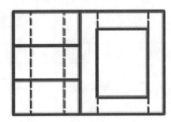

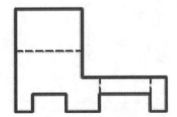 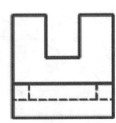

© 2010 Cengage Learning. All Rights Reserved. May not be scanned, copied or duplicated, or posted to a publicly accessible website, in whole or in part.

Problem 1-29

Create the isometric, oblique, or perspective views of the object as directed by your instructor.

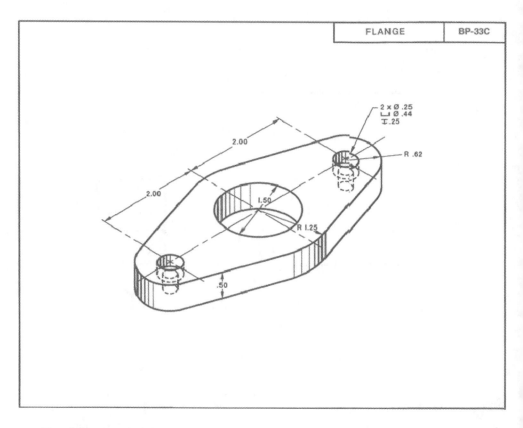

	FLANGE	BP-33C

Problem 1-30

Create the isometric, oblique, or perspective views of the object as directed by your instructor.

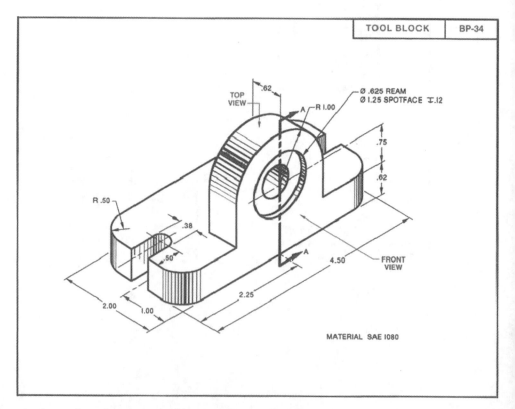

	TOOL BLOCK	BP-34

© 2010 Cengage Learning. All Rights Reserved. May not be scanned, copied or duplicated, or posted to a publicly accessible website, in whole or in part.

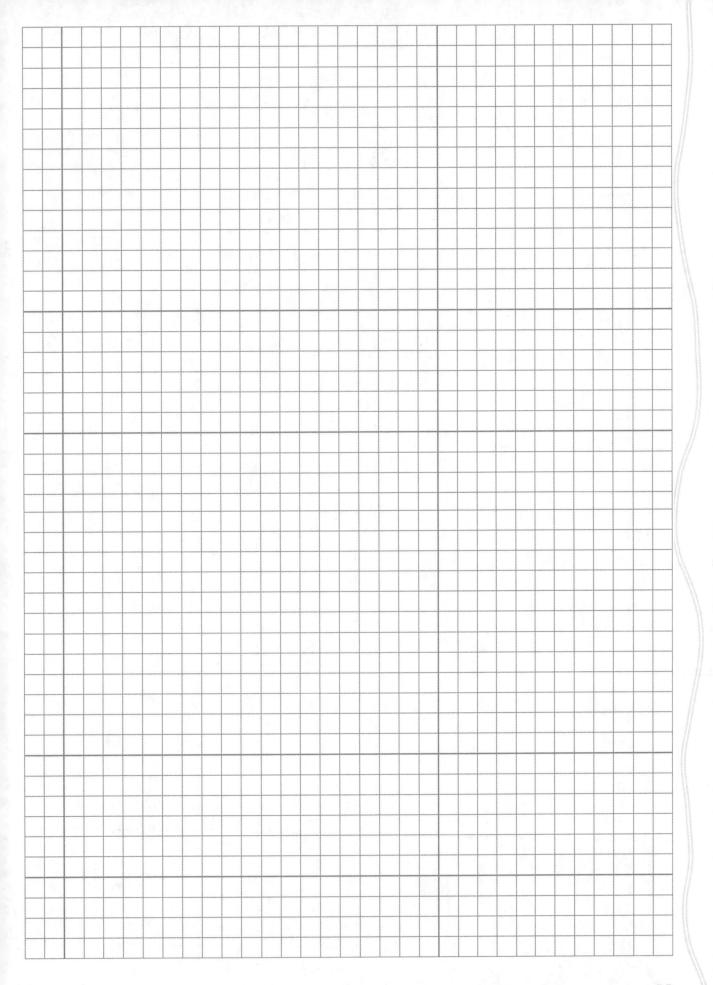

© 2010 Cengage Learning. All Rights Reserved. May not be scanned, copied or duplicated, or posted to a publicly accessible website, in whole or in part.

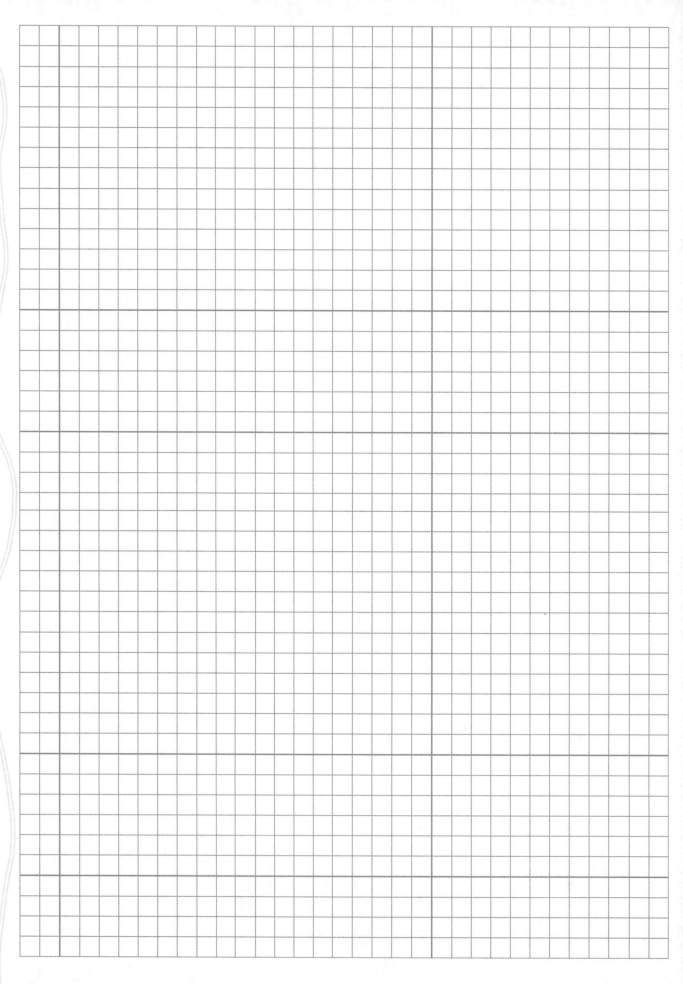

© 2010 Cengage Learning. All Rights Reserved. May not be scanned, copied or duplicated, or posted to a publicly accessible website, in whole or in part.

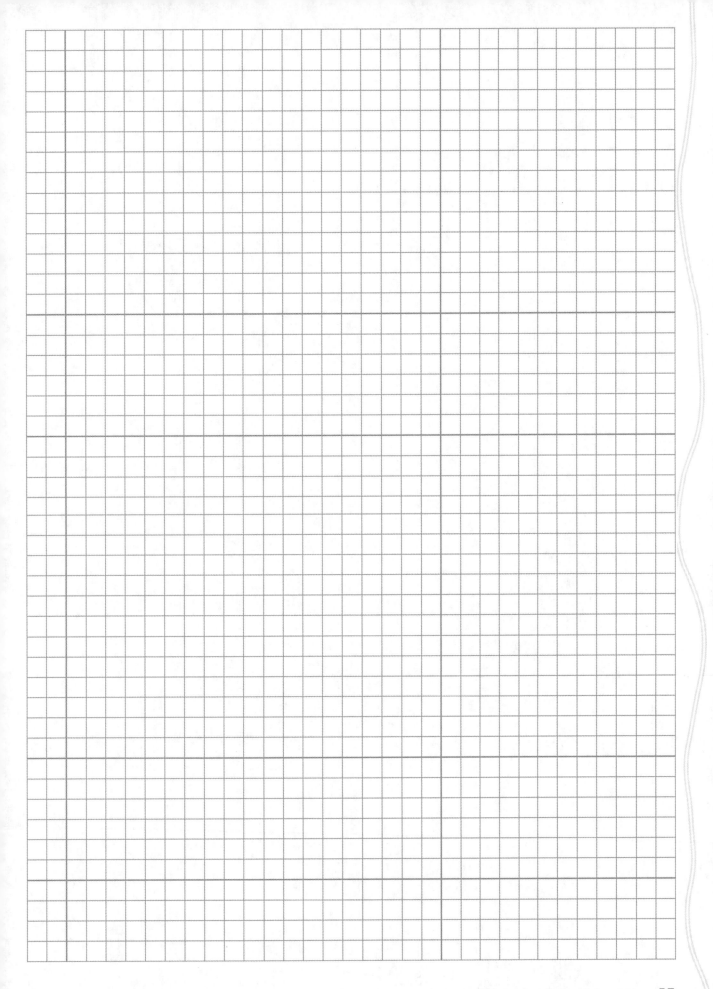

© 2010 Cengage Learning. All Rights Reserved. May not be scanned, copied or duplicated, or posted to a publicly accessible website, in whole or in part.

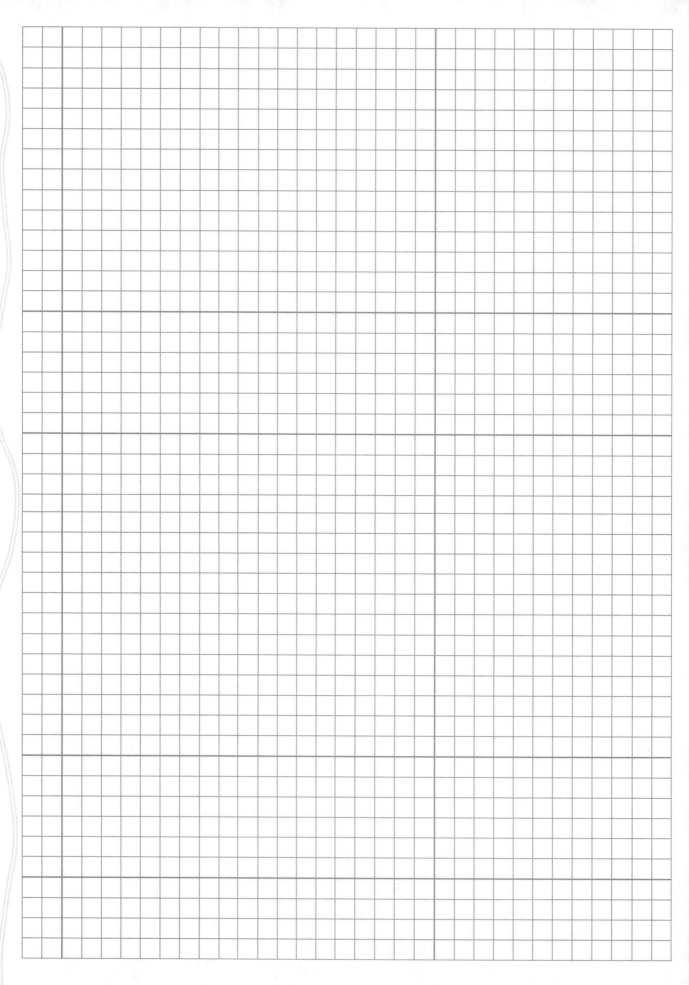

© 2010 Cengage Learning. All Rights Reserved. May not be scanned, copied or duplicated, or posted to a publicly accessible website, in whole or in part.

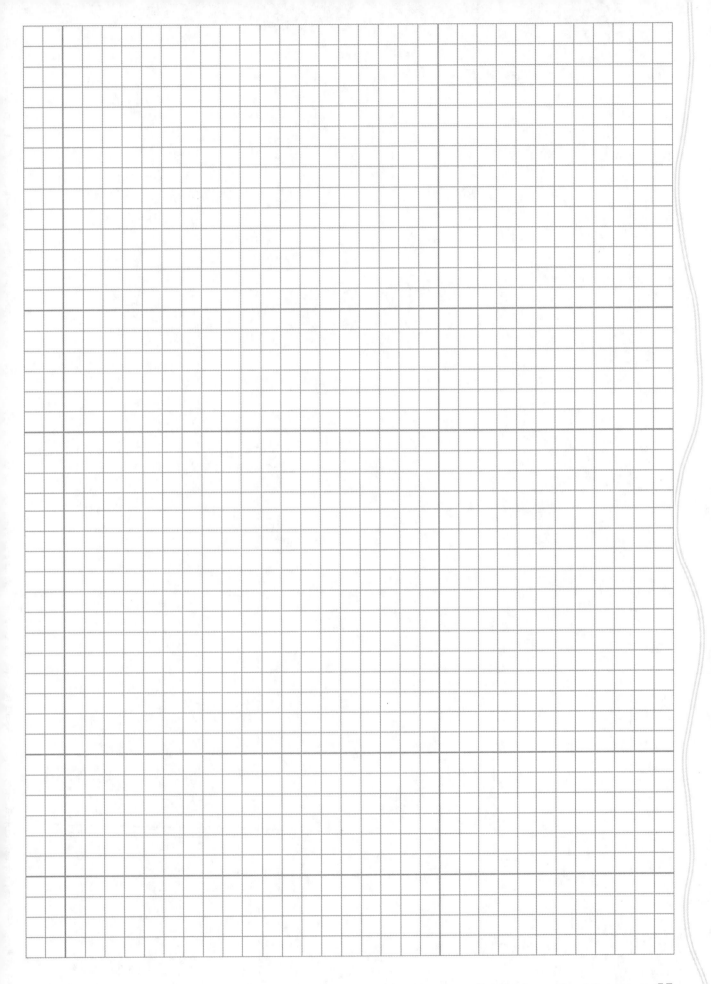

© 2010 Cengage Learning. All Rights Reserved. May not be scanned, copied or duplicated, or posted to a publicly accessible website, in whole or in part.

© 2010 Cengage Learning. All Rights Reserved. May not be scanned, copied or duplicated, or posted to a publicly accessible website, in whole or in part.

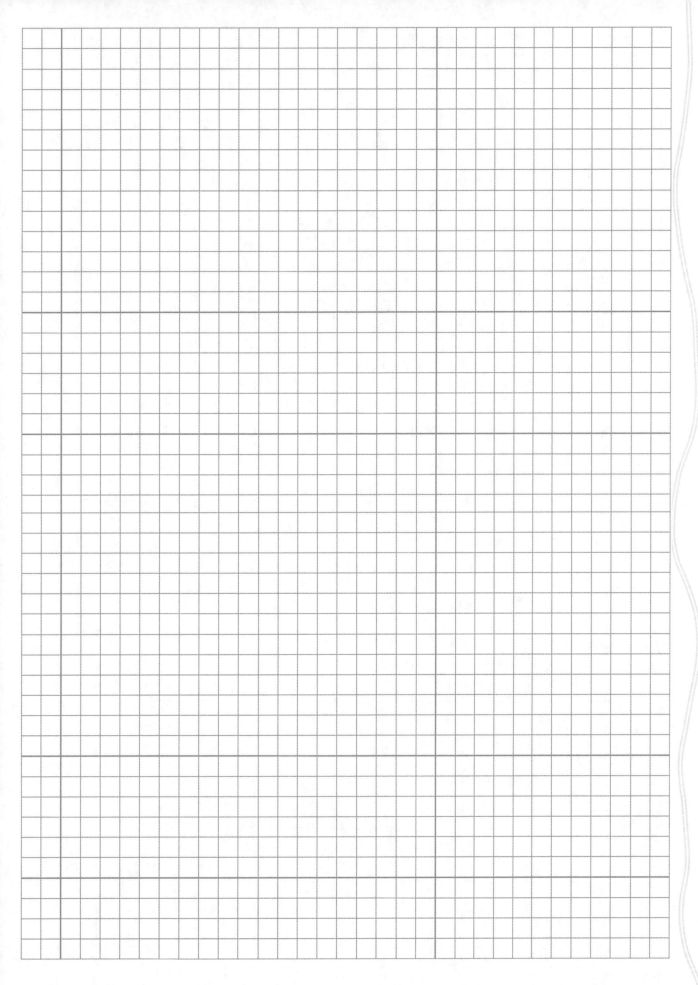

© 2010 Cengage Learning. All Rights Reserved. May not be scanned, copied or duplicated, or posted to a publicly accessible website, in whole or in part.

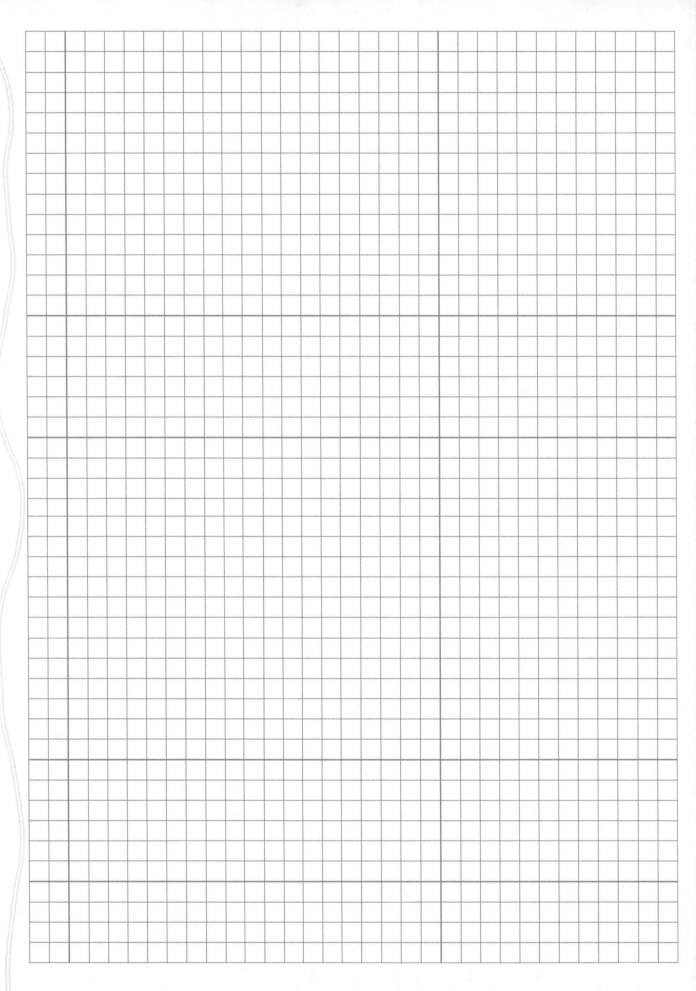

© 2010 Cengage Learning. All Rights Reserved. May not be scanned, copied or duplicated, or posted to a publicly accessible website, in whole or in part.

© 2010 Cengage Learning. All Rights Reserved. May not be scanned, copied or duplicated, or posted to a publicly accessible website, in whole or in part.

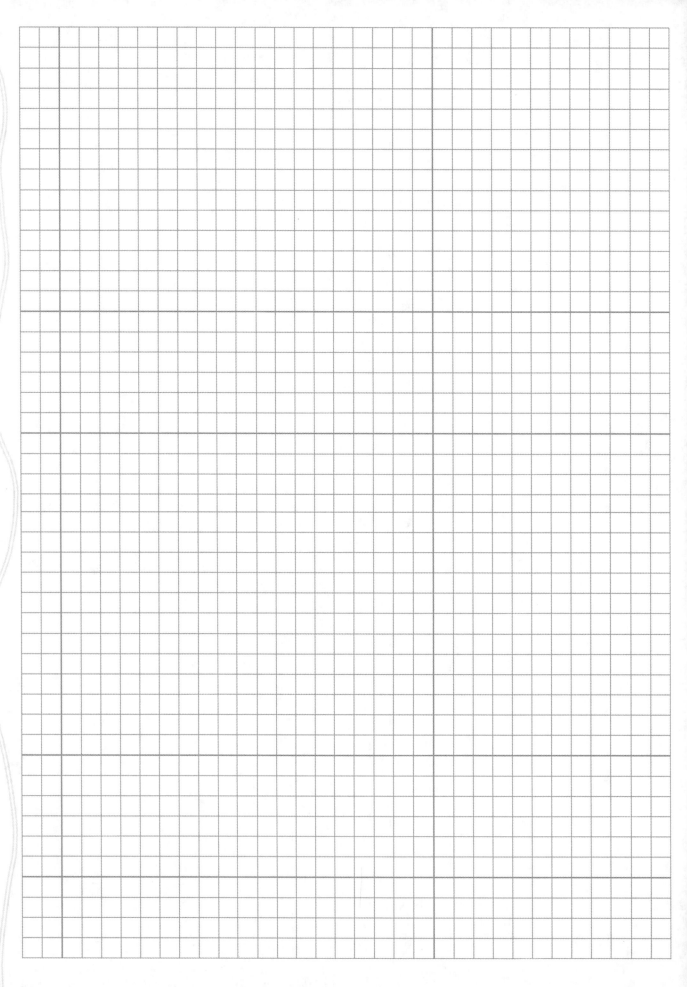

© 2010 Cengage Learning. All Rights Reserved. May not be scanned, copied or duplicated, or posted to a publicly accessible website, in whole or in part.

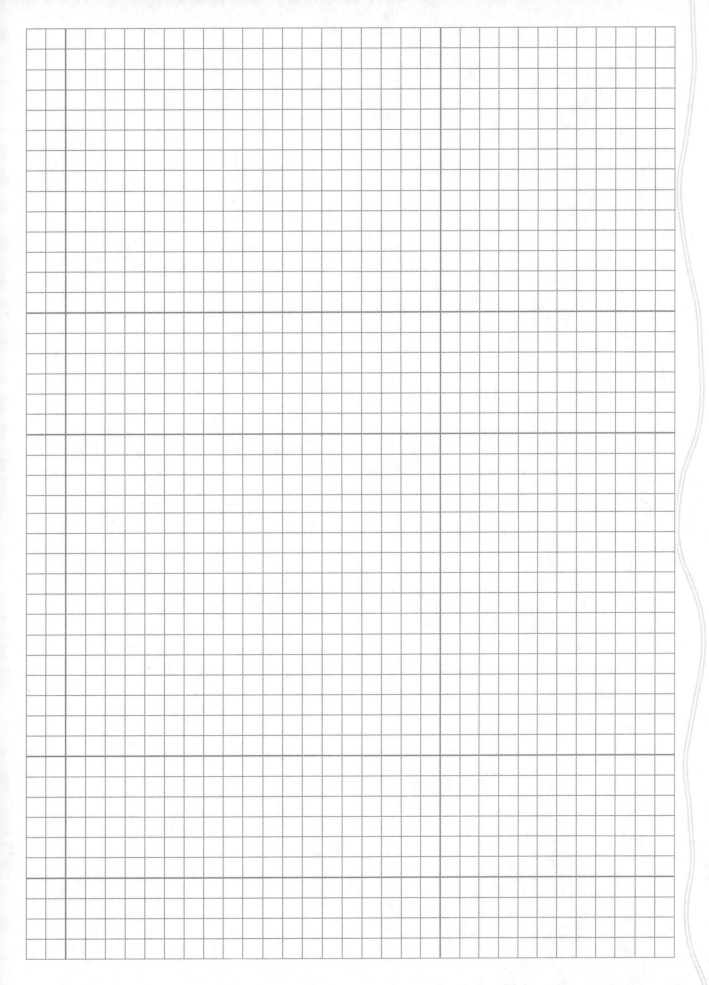

© 2010 Cengage Learning. All Rights Reserved. May not be scanned, copied or duplicated, or posted to a publicly accessible website, in whole or in part.

© 2010 Cengage Learning. All Rights Reserved. May not be scanned, copied or duplicated, or posted to a publicly accessible website, in whole or in part.

© 2010 Cengage Learning. All Rights Reserved. May not be scanned, copied or duplicated, or posted to a publicly accessible website, in whole or in part.

© 2010 Cengage Learning. All Rights Reserved. May not be scanned, copied or duplicated, or posted to a publicly accessible website, in whole or in part.

© 2010 Cengage Learning. All Rights Reserved. May not be scanned, copied or duplicated, or posted to a publicly accessible website, in whole or in part.

© 2010 Cengage Learning. All Rights Reserved. May not be scanned, copied or duplicated, or posted to a publicly accessible website, in whole or in part.

© 2010 Cengage Learning. All Rights Reserved. May not be scanned, copied or duplicated, or posted to a publicly accessible website, in whole or in part.

© 2010 Cengage Learning. All Rights Reserved. May not be scanned, copied or duplicated, or posted to a publicly accessible website, in whole or in part.

© 2010 Cengage Learning. All Rights Reserved. May not be scanned, copied or duplicated, or posted to a publicly accessible website, in whole or in part.

© 2010 Cengage Learning. All Rights Reserved. May not be scanned, copied or duplicated, or posted to a publicly accessible website, in whole or in part.

© 2010 Cengage Learning. All Rights Reserved. May not be scanned, copied or duplicated, or posted to a publicly accessible website, in whole or in part.

© 2010 Cengage Learning. All Rights Reserved. May not be scanned, copied or duplicated, or posted to a publicly accessible website, in whole or in part.

© 2010 Cengage Learning. All Rights Reserved. May not be scanned, copied or duplicated, or posted to a publicly accessible website, in whole or in part.

© 2010 Cengage Learning. All Rights Reserved. May not be scanned, copied or duplicated, or posted to a publicly accessible website, in whole or in part.

© 2010 Cengage Learning. All Rights Reserved. May not be scanned, copied or duplicated, or posted to a publicly accessible website, in whole or in part.

© 2010 Cengage Learning. All Rights Reserved. May not be scanned, copied or duplicated, or posted to a publicly accessible website, in whole or in part.

© 2010 Cengage Learning. All Rights Reserved. May not be scanned, copied or duplicated, or posted to a publicly accessible website, in whole or in part.

© 2010 Cengage Learning. All Rights Reserved. May not be scanned, copied or duplicated, or posted to a publicly accessible website, in whole or in part.

© 2010 Cengage Learning. All Rights Reserved. May not be scanned, copied or duplicated, or posted to a publicly accessible website, in whole or in part.

© 2010 Cengage Learning. All Rights Reserved. May not be scanned, copied or duplicated, or posted to a publicly accessible website, in whole or in part.

UNIT 2
Print Reading and Parametric Modeling

Skills List

After completing the activities in this unit, you should be able to:

- Read engineering drawings

- Understand the math behind the drawing

- Master Inventor basics in:
  o Part creation
  o Drawing (print) development
  o Assemblies
    o Assembled
    o Exploded

- Assemble a drawing sheet packet

© 2010 Cengage Learning. All Rights Reserved. May not be scanned, copied or duplicated, or posted to a publicly accessible website, in whole or in part.

SECTION 1
Print Reading

BACKGROUND

Reading Engineering Drawings

Reading engineering drawings can be very confusing from how to find that missing dimension to interpreting dimensions needed to create a 3-D model. The following will help you with the basics of print reading skills. Additionally, there are some exercises that will help you view and record answers in reading prints.

TIP SHEET

Details, Details, Details

Reading engineering drawings is more about the details; most people reading prints gloss over the details only to look at shapes. The details are in the dimensions, notes, choice of views, and view layouts. Take your time and look over the whole print before making any decisions or judgments on the design.

© 2010 Cengage Learning. All Rights Reserved. May not be scanned, copied or duplicated, or posted to a publicly accessible website, in whole or in part.

EXERCISE 2.1 PRINT READING

Objective

Accomplish the following print reading exercises to gain a focus on reading engineering drawings.

Procedure

Read and fill out the corresponding questions about the engineering drawings.

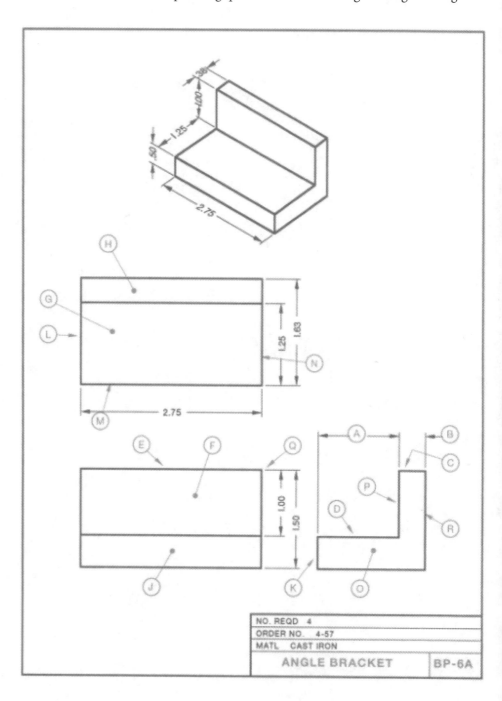

© 2010 Cengage Learning. All Rights Reserved. May not be scanned, copied or duplicated, or posted to a publicly accessible website, in whole or in part.

ANGLE BRACKET (BP-6A)

Student's name _____

The following questions relate to the assembly drawing

1. How many Angle Brackets are required?

2. Name the material specified for the Angle Bracket.

3. Rate the order number of the Bracket.

4. What is the overall width (length) of the Bracket?

5. What is the overall height?

6. What is the overall depth?

7. What is dimension Ⓐ?

8. What is dimension Ⓑ?

9. What surface in the top view is represented by the line Ⓒ in the right side view?

10. Name the three views that are used to describe the shape and size of the part?

11. What surface in the top view is represented by the line Ⓓ in the right side view?

12. What line in the right side view represents surface Ⓕ in the front view?

13. What line in the right side view represents surface Ⓙ in the front view?

14. What line in the top view represents surface Ⓞ in the right side view?

15. What line in the front view represents surface Ⓗ in the top view?

16. What line in the right side view represents surface Ⓗ in the top view?

17. What kind of lines are Ⓔ Ⓛ Ⓒ Ⓓ and Ⓚ?

18. What kinds of lines are Ⓐ and Ⓑ?

19. What encircled letter denotes an extension line?

20. What encircled letter in the front view denotes an object line?

1. _____

2. _____

3. _____

4. _____

5. _____

6. _____

7. _____

8. _____

9. _____

10. _____

11. _____

12. _____

13. _____

14. _____

15. _____

16. _____

17. _____

18. _____

19. _____

20. _____

© 2010 Cengage Learning. All Rights Reserved. May not be scanned, copied or duplicated, or posted to a publicly accessible website, in whole or in part.

EXERCISE 2.2   PRINT READING

Objective

Accomplish the following print reading exercise to gain a focus on reading engineering drawings.

Procedure

Read and fill out the corresponding questions about the engineering drawings.

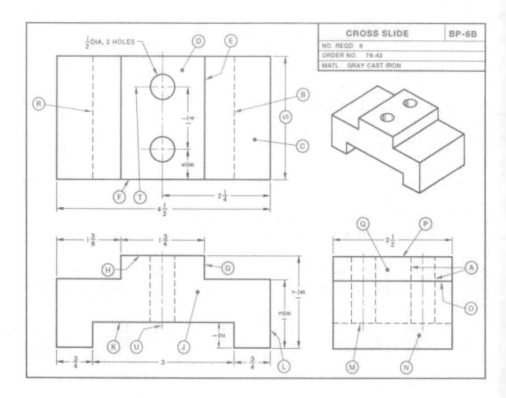

© 2010 Cengage Learning. All Rights Reserved. May not be scanned, copied or duplicated, or posted to a publicly accessible website, in whole or in part.

CROSS SLIDE (BP-6B)

1. What material is used for the Cross Slide?

2. How many pieces are required?

3. What is the overall width (length) of the Cross Slide?

4. What is the order number?

5. What is the overall height of the Cross Slide?

6. What are the lines marked Ⓐ and Ⓑ called?

7. What do the lines marked Ⓐ represent?

8. What two lines in the top view represent the slot shown in the front view?

9. What line in the right side view represents the slot shown in the front view?

10. What line in the front view represents surface Ⓠ in the right side view?

11. What line in the front view represents surface Ⓓ in the top view?

12. What line in the top view represents surface Ⓙ in the front view?

13. What line in the side view represents surface Ⓓ in the top view?

14. What is the diameter of the holes?

15. What is the center-to-center dimension of the holes?

16. How far is the center of the first hole from the front surface of the side?

17. Are the holes drilled all the way through the slide?

18. What is the width of the slot shown in the front view?

19. What is the height of the slot?

20. Determine dimension Ⓢ.

21. What is the width of the projection at the top of the slide?

22. How high is the projection?

23. What kind of line is Ⓜ?

24. What kind of line is used at Ⓞ and Ⓟ?

Student's name _____

1. _____

2. _____

3. _____

4. _____

5. _____

6. _____

7. _____

8. _____

9. _____

10. _____

11. _____

12. _____

13. _____

14. _____

15. _____

16. _____

17. _____

18. _____

19. _____

20. _____

21. _____

22. _____

23. _____

24. _____

© 2010 Cengage Learning. All Rights Reserved. May not be scanned, copied or duplicated, or posted to a publicly accessible website, in whole or in part.

Objective

Accomplish the following print reading exercises to gain a focus on reading engineering drawings.

Procedure

Read and fill out the corresponding questions about the engineering drawings.

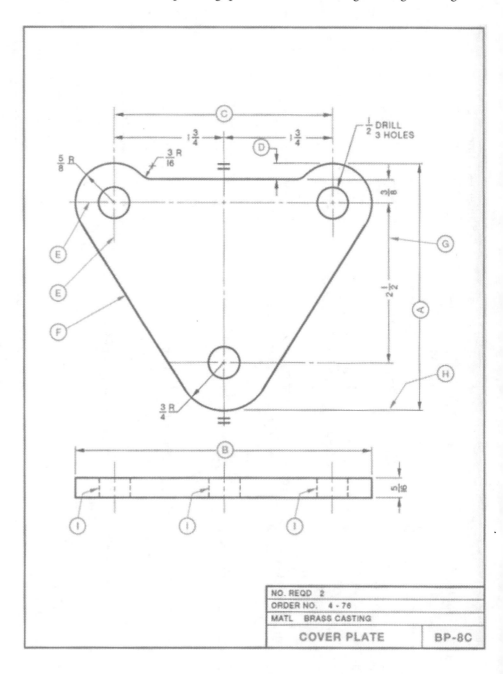

© 2010 Cengage Learning. All Rights Reserved. May not be scanned, copied or duplicated, or posted to a publicly accessible website, in whole or in part.

COVER PLATE (BP-BC)

1. Name the material specified for the part.

2. Name the two views used to describe the part.

3. Identify the kind of line indicated by each of the following encircled letters.

   Ⓔ

   Ⓕ

   Ⓖ

   Ⓗ

   Ⓘ

4. What is the overall depth Ⓐ?

5. What is the overall length Ⓑ?

6. How many holes are to be drilled?

7. What is the thickness (height) of the plate?

8. What is the diameter of the holes?

9. What is the distance between the center of one of the two upper holes and the center line of the plate?

10. Give the center distance Ⓒ of the two upper holes.

11. What is the radius that forms the two upper rounds of the Plate?

12. What radius forms the lower part of the Plate?

13. What kind of line is drawn through the center of the Plate?

14. Determine distance Ⓓ.

15. How much stock is left between the edge of one of the upper holes and the outside of the piece?

Student's name _____

1. _____

2. _____

3.
   E _____

   F _____

   G _____

   H _____

   I _____

4. _____

5. _____

6. _____

7. _____

8. _____

9. _____

10. _____

11. _____

12. _____

13. _____

14. _____

15. _____

© 2010 Cengage Learning. All Rights Reserved. May not be scanned, copied or duplicated, or posted to a publicly accessible website, in whole or in part.

EXERCISE 2.4   PRINT READING

Objective

Accomplish the following print reading exercises to gain a focus on reading engineering drawings.

Procedure

Read and fill out the corresponding questions about the engineering drawings.

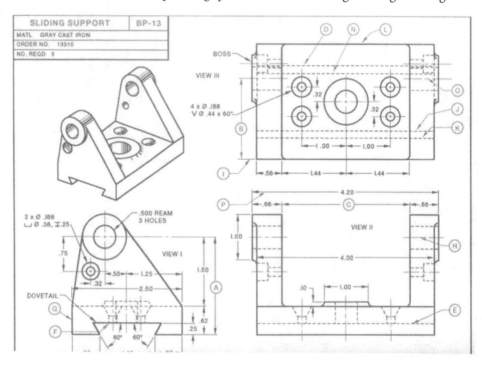

© 2010 Cengage Learning. All Rights Reserved. May not be scanned, copied or duplicated, or posted to a publicly accessible website, in whole or in part.

SLIDING SUPPORT (BP-13)

1. Name the view I which shows the shape of the dovetail.

2. Name view III in which the bottom pad appears as a circle.

3. Name view II

4. Name the kind of line shown at **E**.

5. What surface in view I is represented by line **E**?

6. Name the kind of line shown at **G**.

7. What line in view III represents surfaces **G**?

8. Name the kind of line shown at **H**.

9. What line in the view III represents the line **H**?

10. Name the kind of line shown at **I**.

11. Name the kind of line shown at **J**.

12. What lines in the top view represent the dovetail?

13. What does the line **E** in the front view represent?

14. Determine height **A**.

15. How many bosses are shown on the uprights?

16. What is the outside diameter of the boss?

17. Determine dimension **B**.

18. How far off from the center of the support is the center of the two holes in the bosses of the uprights?

19. Give the dimensions for the counterbored holes.

20. What dimensions are given for the countersunk holes?

21. Give the dimensions for the reamed holes.

22. What is the dimension **C**?

23. How wide is the opening in the dovetail?

24. How deep is the dovetail machined?

25. What is the angle to the horizontal at which the dovetail is cut?

Student's name _____

1. _____
2. _____
3. _____
4. _____
5. _____
6. _____
7. _____
8. _____
9. _____
10. _____
11. _____
12. _____
13. _____
14. _____
15. _____
16. _____
17. _____
18. _____
19. _____
20. _____
21. _____
22. _____
23. _____
24. _____
25. _____

© 2010 Cengage Learning. All Rights Reserved. May not be scanned, copied or duplicated, or posted to a publicly accessible website, in whole or in part.

EXERCISE 2.5 PRINT READING

Objective

Accomplish the following print reading exercises to gain a focus on reading engineering drawings.

Procedure

Read and fill out the corresponding questions about the engineering drawings.

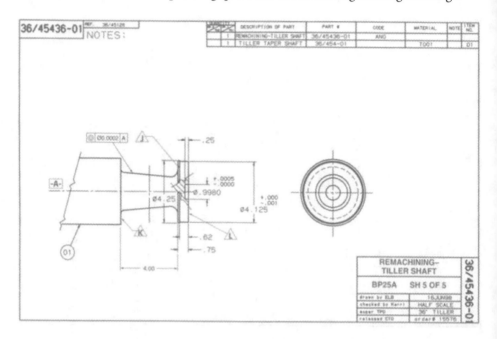

© 2010 Cengage Learning. All Rights Reserved. May not be scanned, copied or duplicated, or posted to a publicly accessible website, in whole or in part.

Tiller Taper Shaft Assembly (BP-25A)

Student's name _____

The following questions relate to the assembly drawing

1. Print the name of the following parts:

   Ⓐ , Ⓑ , Ⓒ , Ⓓ

2. Give the drawing number for the tiller taper shaft assembly.

3. How many detail and assembly drawings make up the tiller taper shaft working drawing?

4. How many parts are needed to make a tiller taper shaft assembly?

5. How many of these parts are commercial or purchased parts?

1. Ⓐ _____

   Ⓑ _____

   Ⓒ _____

   Ⓓ _____

2. _____

3. _____

4. _____

5. _____

The following questions relate to the bearing.

1. Tell why the section lines of (04) go in two different directions.

2. Give the minimum taper angle of the bearing.

3. What is the part number of the bearing?

4. How many pieces make up the bearing?

5. What is the basic height and thickness of the bearing?

1. _____

2. _____

3. _____

4. _____

5. _____

The following questions relate to the ring.

1. What is the thickness of the ring?

2. What is the basic OD of the ring?

3. What is the minimum hole size of the ring when using the upper limit of the shaft & lower limit of the ring?

4. Explain the meaning of "1.0000 Hold"

5. Explain why the 2.00 REF is given.

1. _____

2. _____

3. _____

4. _____

5. _____

© 2010 Cengage Learning. All Rights Reserved. May not be scanned, copied or duplicated, or posted to a publicly accessible website, in whole or in part.

SECTION 2
Math Basics behind the Drawing

BACKGROUND

Drawings and Math

Drawing and design is all math. Yes, that sounds scary but in reality it is manageable. It is a combination of geometry, basic math, and at times a little algebra. Starting with a basic drawing there is always some addition or subtraction involved. The following is a basic drawing with several dimensions. As you can see, there is one dimension that is missing but an overall dimension is provided. Add up the small dimensions and subtract them from the overall dimension for the missing value. See Figure 2-1.

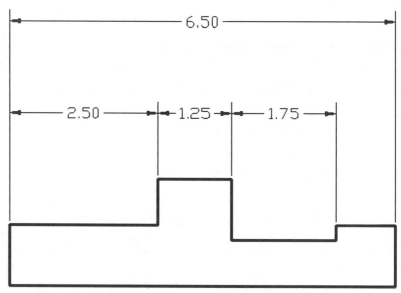

FIGURE 2-1 *Continuous dimensioning practices.*

BACKGROUND

Baseline Dimensions

Baseline dimensions use a common surface as a starting point to each of the features. This style eliminates the math calculations for the missing incremental dimension like the previous example. See Figure 2-2.

© 2010 Cengage Learning. All Rights Reserved. May not be scanned, copied or duplicated, or posted to a publicly accessible website, in whole or in part.

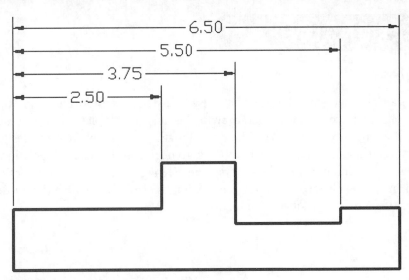

FIGURE 2-2 *Baseline dimensioning practices.*

BACKGROUND

Figuring It Out. Ordinate Dimensions

There are also varied dimensioning styles that drawings can use. The next drawing uses an ordinate dimensioning system. It is based on a common zero location, not specifically an edge of a drawing.

In ordinate dimensioning the zero location provides the starting point in dimensioning all features and limits on an object. The .50-inch-diameter holes on the far right side of the object are located 8 inches from the left-side zero location and 3.0 and 4.5 inches respectively from the bottom of the drawing. See Figure 2-3.

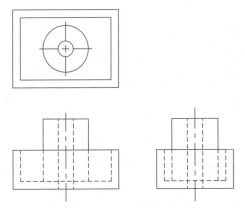

FIGURE 2-3 *Ordinate dimensioning practices.*

© 2010 Cengage Learning. All Rights Reserved. May not be scanned, copied or duplicated, or posted to a publicly accessible website, in whole or in part.

Angles

Angles are another geometry area that gets confusing. It is important to predict approximate angles, especially when drawing because angles play an important part in the drawing process in CAD. The typical mathematical, mechanical design/architectural design angular format has the zero degree in the 3 o'clock position. The 90-degree position is at 12 o'clock, and so on. See Figure 2-4.

For mapping designs, the 0-degree position is at 12 o'clock with the 90-degree angle at the 3 o'clock position. See Figure 2-5.

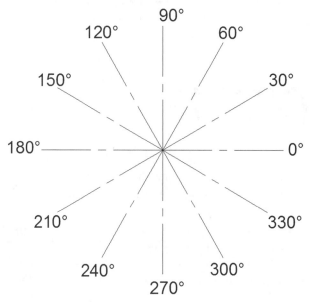

FIGURE 2-4 *Math-, science-, and drafting-based angles.*

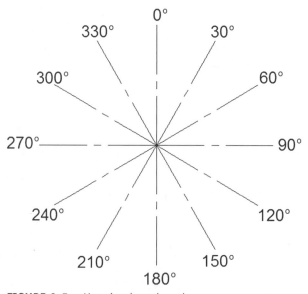

FIGURE 2-5 *Mapping-based angles.*

© 2010 Cengage Learning. All Rights Reserved. May not be scanned, copied or duplicated, or posted to a publicly accessible website, in whole or in part.

Cartesian Coordinate System

Positions of object features are denoted by an X, Y, and Z position in space. These are called *coordinates* and they are placed using the Cartesian coordinate system.

For 2-D work we focus on the first two numbers, X and Y (X is horizontal and Y is vertical). Check any icons that represent X and Y to make sure of the direction prior to creating any object shapes to make sure the orientation of the object is appropriate. Inventor and other parametric modelers allow for direct coordinate entry for initial accuracy in the drawing sketch creation process.

The X–Y coordinate system with data points (2,5) x = 2 and y = 5. See Figure 2-6.

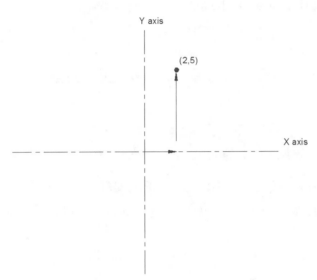

FIGURE 2-6 *Cartesian coordinate system.*

Confusion in Axis Land

Do not confuse the Cartesian coordinate system with angular measurement. Both use axis layouts, but coordinates represent positional location and angular measurements represent angles. It is a common confusion for new designers.

The 3-D coordinate system is color based since it would be hard to make out the axis quickly in a drawing.

© 2010 Cengage Learning. All Rights Reserved. May not be scanned, copied or duplicated, or posted to a publicly accessible website, in whole or in part.

3-D Colors Are Standard

The colors used in 3-D axis locations are standardized throughout all CAD and design software tools. The colors may not show up until an object is shaded. See Figure 2-7.

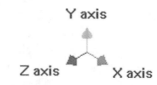

FIGURE 2-7 *3-D axis showing XYZ coordinate system.*

- The X axis represents the red leg
- The Y axis represents the green leg
- The Z axis represents the blue leg

BACKGROUND

Sample Scales to Use for Measurement

This ruler page can be cut out and used for measurement of worksheets or objects.

© 2010 Cengage Learning. All Rights Reserved. May not be scanned, copied or duplicated, or posted to a publicly accessible website, in whole or in part.

10

0   1   2   3   4   5   6   7   8   9

1/16

0   1   2   3   4   5   6   7   8   9

Metric

0   1   2   3   4   5   6   7   8   9   10   11   12   13   14   15   16   17   18   19   20   21   22   23   24   25

© 2010 Cengage Learning. All Rights Reserved. May not be scanned, copied or duplicated, or posted to a publicly accessible website, in whole or in part.

EXERCISE 2.6   MEASUREMENT OF OBJECTS

Objective

Obtain accurate measurements of the objects.

Materials

☐ Scale or ruler page from workbook

☐ Pencil

Procedure

Measure and write the dimension for the indicated scale.

1. Measure the line below using 1/16th scale and metric:

   _____

2. Measure the line below using 1/10th scale and metric:

   _____

3. Measure sides of the rectangle below using 1/16th scale and metric:

4. Measure the sides of this triangle using metric and 1/10th scale:

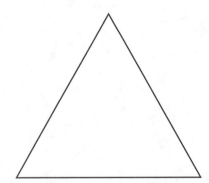

© 2010 Cengage Learning. All Rights Reserved. May not be scanned, copied or duplicated, or posted to a publicly accessible website, in whole or in part.

SECTION 3
Geometric Shapes

Drawing Is Geometry

Every time you make a sketch or use CAD software, you are actually "doing" geometry at some level. The lines and shapes you make can be related to a concept in geometry. Are the lines perpendicular? Is the shape a rectangle or a square? Once you have made and identified shapes you can then calculate important information such as area or perimeter. If you go beyond a 2-D shape to a 3-D form or solid, you can then calculate even more information such as mass and volume. This information is important to engineers for many reasons. Knowing how much something weighs or how much paint is needed to paint all of the surfaces are important considerations. In this section we are going to learn about geometric shapes and how to calculate some of the properties associated with the shapes.

A geometric shape describes the 2- or 3-D contours that characterize an object, or simply put, it's the outer form of an object or figure. Let's identify some common 2-D common shapes. See Figure 2-8.

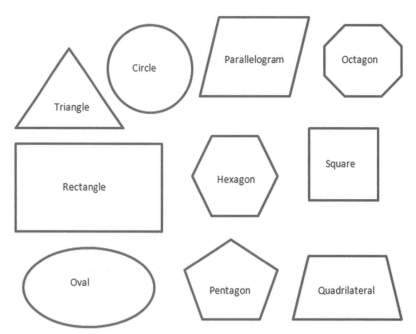

FIGURE 2-8 *Various geometric shapes.*

- Circles—A perfectly round shape.
- Squares—Shape with four sides that are of equal length and at right angles to each other.
- Ellipses—Also called an oval, an ellipse is a flattened-out circle.

© 2010 Cengage Learning. All Rights Reserved. May not be scanned, copied or duplicated, or posted to a publicly accessible website, in whole or in part.

- Rectangles—Shape that has four sides and each side is at right angles to each other.
- Triangles—Shape with three sides.
- Hexagons—Shape with six sides.
- Octagons—Shape with eight sides.
- Pentagons—Shape with five sides.
- Quadrilaterals—A four-sided polygon with four angles.
- Parallelograms—A quadrilateral with two sets of parallel sides.
- Polygon—A closed shape with a minimum of three sides.

TIP SHEET

How Many Sides Does an Octagon Have?

Are you having trouble figuring out the difference between an octagon and a quadrilateral? Identify the prefix of the word.

- Tri = Three
- Hex = Six
- Oct = Eight
- Deca = Ten
- Poly = Many
- Penta = Five
- Para = Beside
- Quad = Four

Who thinks English class is not important!

© 2010 Cengage Learning. All Rights Reserved. May not be scanned, copied or duplicated, or posted to a publicly accessible website, in whole or in part.

Objective

Visually identifying shapes is important to help figure out how to create them. Identify the shapes.

Materials

- ☐ Paper
- ☐ Pencil

Procedure

Identify the type of shape for each object below:

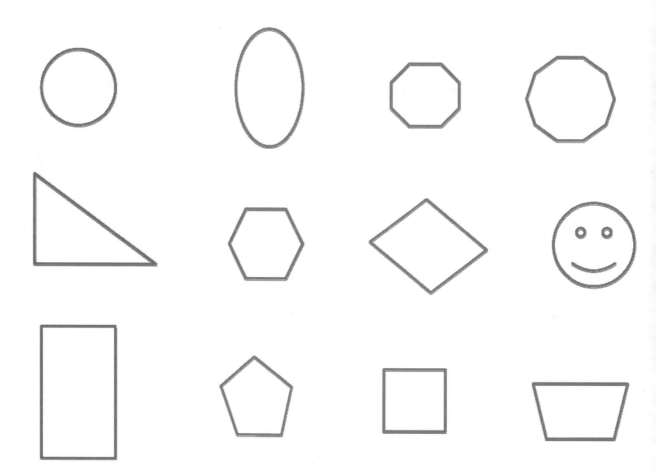

BACKGROUND

Calculating Properties of Shapes

Now that we understand and can identify geometric shapes, let's take a look at how to make some calculations. For a 2-D shape, a common calculation will include area, perimeter, and circumference. Area is defined as the extent or measurement of a surface. You have to know what formula to use depending on the geometric shape. See Figures 2-9 and 2-10.

© 2010 Cengage Learning. All Rights Reserved. May not be scanned, copied or duplicated, or posted to a publicly accessible website, in whole or in part.

Area of Shapes

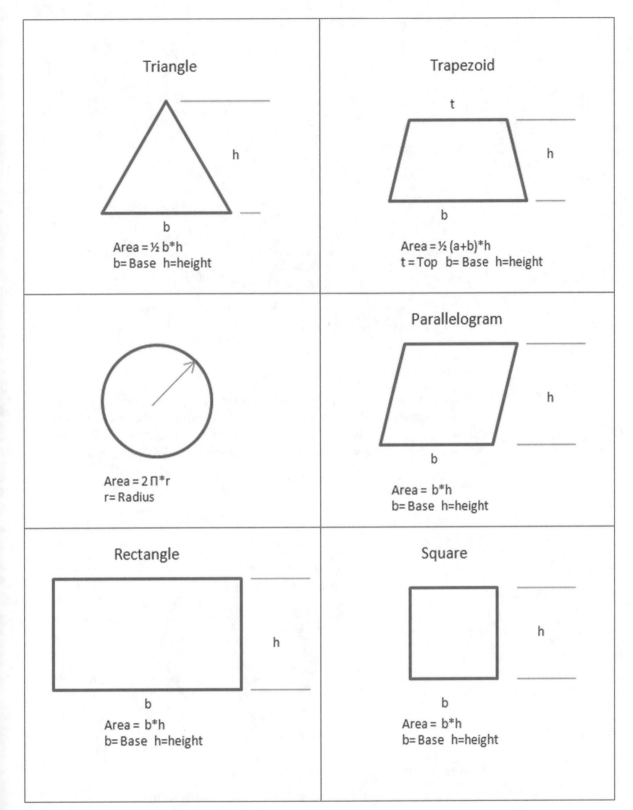

Triangle

Area = ½ b*h
b= Base h=height

Trapezoid

t
h
b

Area = ½ (a+b)*h
t = Top b= Base h=height

Area = 2 Π*r
r= Radius

Parallelogram

h
b

Area = b*h
b= Base h=height

Rectangle

h
b

Area = b*h
b= Base h=height

Square

h
b

Area = b*h
b= Base h=height

FIGURE 2-9 *Calculating the area of shapes.*

© 2010 Cengage Learning. All Rights Reserved. May not be scanned, copied or duplicated, or posted to a publicly accessible website, in whole or in part.

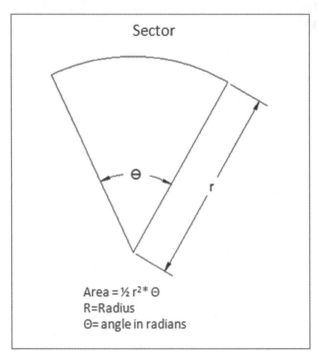

FIGURE 2-10 *Calculating the area of shapes.*

Area of a Circle

Archimedes was an ancient Greek who was the first to calculate the ratio between a circle's diameter and its circumference (distance around the circle); this ratio is known as *pi*. For every unit of diameter distance, the circumference will be 3.14 units. See Figure 2-11. The formula used to calculate the area of a circle is as follows:

$$A = \pi r^2$$

A = area
π = 3.14
r = radius

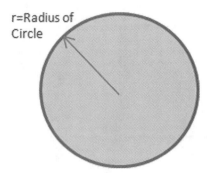

FIGURE 2-11 *Area of a circle.*

Area of a Rectangle

To calculate the area of rectangle, we use the following formula:

$$A = w \times h$$

A = area
w = width
h = height

Example: Note that answers are in units squared. See Figure 2-12.

© 2010 Cengage Learning. All Rights Reserved. May not be scanned, copied or duplicated, or posted to a publicly accessible website, in whole or in part.

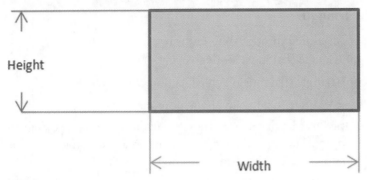

FIGURE 2-12 *Area of a rectangle.*

Area of a Square

To calculate the area of a square we use the following formula. Remember in a square all the sides are equal length. See Figure 2-13.

$$A = side^2$$

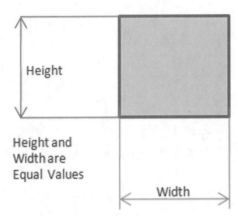

FIGURE 2-13 *Area of a square. Note that answers are in units squared.*

Area of an Ellipse

Remember an ellipse is a geometric shape that looks like a stretched-out circle, so we have to use a formula where we know the length of the major and minor axes. See Figure 2-14.

$$Area = \pi \times a \times b$$

a = major axis length
b = minor axis length

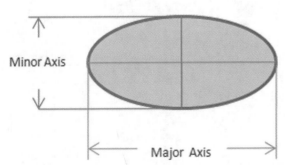

FIGURE 2-14 *Area of an ellipse. Note that answers are in units squared.*

© 2010 Cengage Learning. All Rights Reserved. May not be scanned, copied or duplicated, or posted to a publicly accessible website, in whole or in part.

Area of a Triangle

Triangles are classified three in different ways:

- Right triangle (having a 90-degree angle)
- Acute triangle (having an angle less than 90 degrees)
- Obtuse triangle (having an angle more than 90 degrees)

The sum of the three angles in any triangle will always equal 180 degrees.

Use the appropriate formula to calculate the area of a triangle. To calculate area, you need to know the length of the base and the height.

Right Triangle

- A right triangle has one right (90-degree) angle. See Figure 2-15.

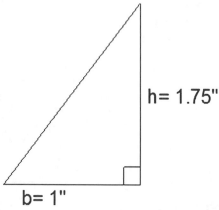

FIGURE 2-15   *Right triangle.*

$$A = 1/2\ b \times h$$

$A = 1/2\ (1\ in) \times (1.5\ in)$
$A = 1/2\ (1.5\ in^2)$
$A = .75\ in^2$

Acute Triangle

- An acute triangle has three acute angles (angles that measure between 0 and 90 degrees). See Figure 2-16.

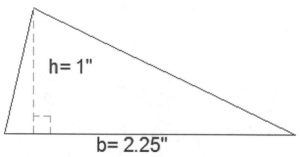

FIGURE 2-16   *Acute triangle.*

$$A = 1/2\ b \times h$$

$A = 1/2\ (1\ in) \times (2.25\ in)$
$A = 1/2\ (2.25\ in^2)$
$A = 1.125\ in^2$

© 2010 Cengage Learning. All Rights Reserved. May not be scanned, copied or duplicated, or posted to a publicly accessible website, in whole or in part.

Obtuse Triangle

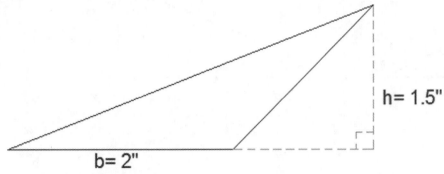

FIGURE 2-17 *Obtuse triangle.*

- An obtuse triangle has one obtuse angle (an angle that measures between 90 and 180 degrees). See Figure 2-17.

$$A = 1/2 \, b \times h$$

A = 1/2 (2 in) × (1.5 in)
A = 1/2 (3 in²)
A = 1.5 in²

EXERCISE 2.8   CALCULATING TRIANGULAR AREAS

Objective

Determine the area of the different triangular shape through calculations.

Materials

☐ Paper
☐ Pencil

Procedure

Calculate the area of each angle.

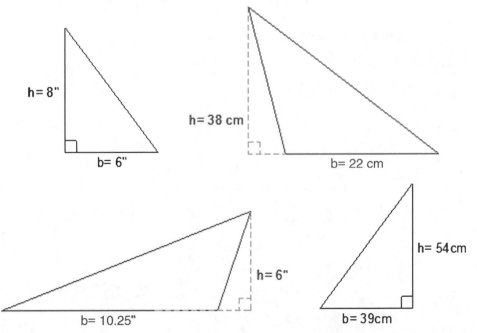

© 2010 Cengage Learning. All Rights Reserved. May not be scanned, copied or duplicated, or posted to a publicly accessible website, in whole or in part.

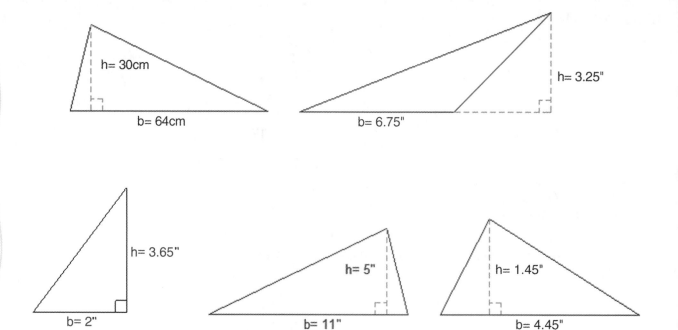

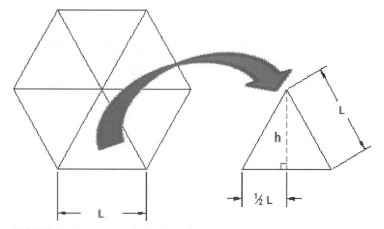

Area of a Hexagon

To find the area of a hexagon you start by first splitting it into six equilateral triangles. See Figure 2-18.

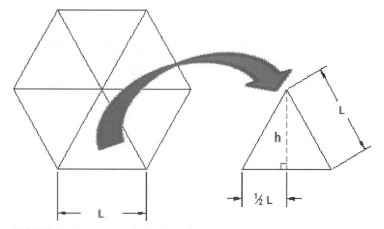

FIGURE 2-18 *Area calculation of a hexagon section.*

- L is the side length.
- H is the height of each triangle.

Find the area for one triangle using the formula A = 1/2 L × h. See Figure 2-19.

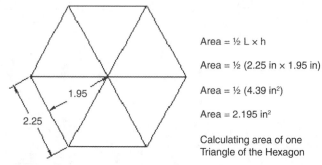

Area = ½ L x h

Area = ½ (2.25 in x 1.95 in)

Area = ½ (4.39 in²)

Area = 2.195 in²

Calculating area of one Triangle of the Hexagon

FIGURE 2-19 *Area of a hexagon.*

© 2010 Cengage Learning. All Rights Reserved. May not be scanned, copied or duplicated, or posted to a publicly accessible website, in whole or in part.

To calculate the hexagon section you need to multiply the answer by six!

Total area of hexagon = 2.195 in² × 6

Total area of hexagon = 13.17 in²

Area of an Octagon

To calculate the area of an octagon, we have to split the octagon into known geometric shapes that we can find the area of. See Figure 2-20.

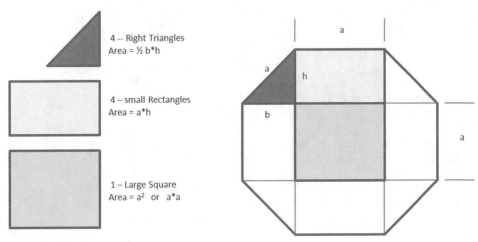

FIGURE 2-20 *Area of an octagon.*

The octagon breaks down to:

- One square
- Four rectangles
- Four triangles
- The h distance can be calculated by the Pythagorean theorem. $h^2 = a^2 + b^2$

Calculate the area of each geometric shape and then add them all together to find the total area of the octagon. See Figure 2-21.

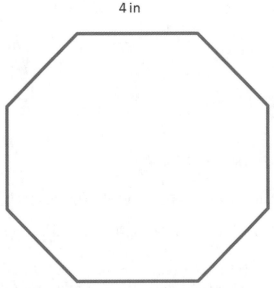

FIGURE 2-21 *Octagon practice problem.*

© 2010 Cengage Learning. All Rights Reserved. May not be scanned, copied or duplicated, or posted to a publicly accessible website, in whole or in part.

Area of a Pentagon

A method to calculate the area of a pentagon (with equal sides) is to divide the shape into five equal segments (from a centerpoint). The next step is to divide the segment into right triangles. Find the area of the right triangles and then add up the total number of right triangles to get the area. See Figure 2-22.

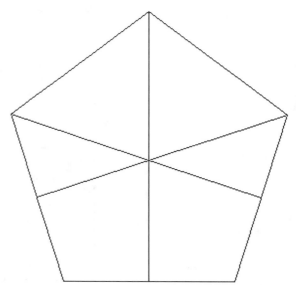

FIGURE 2-22 *Break the pentagon into equal triangle shapes by drawing lines through the center. The two top triangles are the equal-sized triangles.*

To locate the center, draw perpendicular lines starting from the top three corners. Where they cross is the centerpoint. See Figure 2-23.

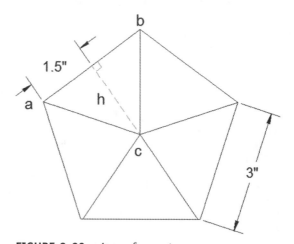

FIGURE 2-23 *Area of a pentagon.*

Pentagons are made up of five equal triangles. Triangle area is A = 1/2 base × height.

In our example the base is 1.5 inches and the height needs to be calculated using basic trigonometry.

With all five triangles being equal we can calculate that $360°/5 = 72°$.

One-half of the angle (since distance h bisects the angle in half) $= 36°$.

$$\text{Tangent (interior angle)} = \frac{\text{Opposite distance}}{\text{Adjacent distance (this is h)}}$$

© 2010 Cengage Learning. All Rights Reserved. May not be scanned, copied or duplicated, or posted to a publicly accessible website, in whole or in part.

$$\text{Tangent } (36°) = .7265$$
$$\text{Tangent } (36°) = \frac{1.5 \text{ in}}{h}$$
$$.7265 = \frac{1.5 \text{ in}}{h}$$
$$h \,(.7625) = 1.5 \text{ in}$$
$$h = 1.967 \text{ in}$$

Single pentagon triangle area
$$\text{Area} = 1/2 \,b \times h$$
$$\text{Area} = 1/2 \,(1.5 \text{ in} \times 1.967 \text{ in})$$
$$\text{Area} = 1.475 \text{ in}^2$$
$$\text{Total pentagon area} = 5 \times 1.475 \text{ in}^2$$
$$\text{Total pentagon area} = 7.375 \text{ in}^2$$

Sample Problem: Calculate the total pentagon area. See Figure 2-24.

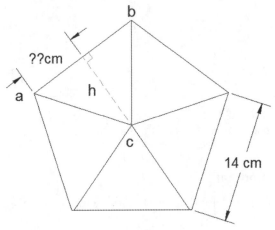

FIGURE 2-24 *Practice problem on area of pentagon.*

Area of a Quadrilateral

To calculate the area of a quadrilateral, divide the geometric shape into two triangles and then add the area of each triangle together. See Figures 2-25 and 2-26.

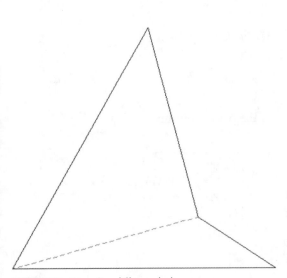

FIGURE 2-25 *Quadrilateral shape.*

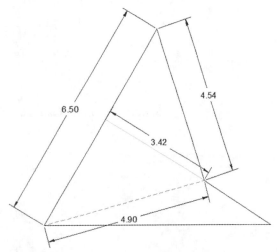

FIGURE 2-26 *Break quadrilateral into two triangles.*

© 2010 Cengage Learning. All Rights Reserved. May not be scanned, copied or duplicated, or posted to a publicly accessible website, in whole or in part.

$$\text{Area of triangle} = 1/2 \ b \times h$$

b = 6.50, h = 3.42
Area = 1/2 (6.5 × 3.42)
Area of one of the quadrilateral triangles = 11.115 in²

Sample Problem: Calculate the total quadrilateral area.

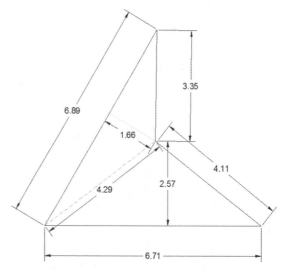

FIGURE 2-27 *Use the dimensions on this quadrilateral to calculate the total area.*

Area of a Parallelogram

To find the area of a parallelogram, use the following formula. See Figure 2-28.

$$A = b \times h$$

A = area
b = base
h = height

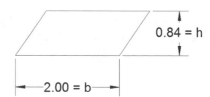

FIGURE 2-28 *Parallelogram area.*

Area = 2 × .84
Area = 1.68 in²

Sample Problem: Calculate the area of the parallelogram. See Figure 2-29.

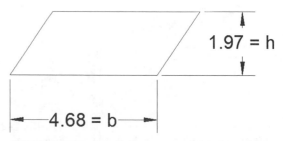

FIGURE 2-29 *Parallelogram area sample problem.*

© 2010 Cengage Learning. All Rights Reserved. May not be scanned, copied or duplicated, or posted to a publicly accessible website, in whole or in part.

EXERCISE 2.9   CALCULATING AREA

Objective
Calculate the area of the object shapes.

Materials

☐ Pencil
☐ Calculator

Procedure
Calculate the area of each shape by using math techniques and equations.

1. Calculate the area of the hexagon.

    Area: _____

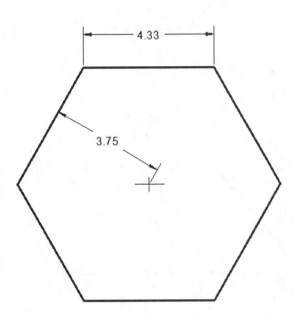

2. Calculate the area of the triangle.

    Area: _____

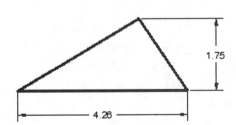

© 2010 Cengage Learning. All Rights Reserved. May not be scanned, copied or duplicated, or posted to a publicly accessible website, in whole or in part.

3. Calculate the area of the hexagon.

   Area: _____

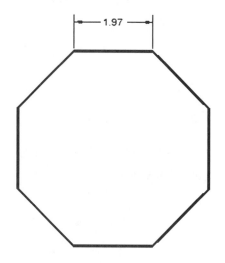

4. Calculate the area of the parallelogram.

   Area: _____

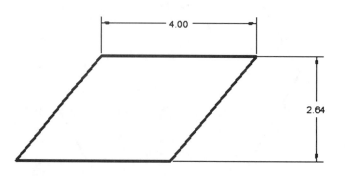

5. Calculate the area of the triangle.

   Area: _____

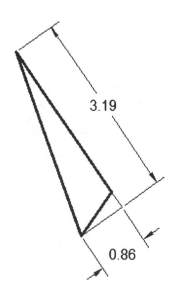

© 2010 Cengage Learning. All Rights Reserved. May not be scanned, copied or duplicated, or posted to a publicly accessible website, in whole or in part.

6. Calculate the area of the pentagon.

   Area: _____

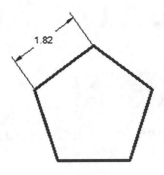

7. Calculate the area of the rectangle.

   Area: _____

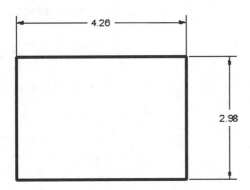

8. Calculate the area of the triangle.

   Area: _____

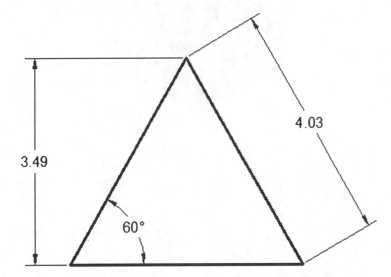

© 2010 Cengage Learning. All Rights Reserved. May not be scanned, copied or duplicated, or posted to a publicly accessible website, in whole or in part.

SECTION 4
Geometric Solids

Calculating Volumes and Surface Areas of 3-D Objects

The difference between a geometric shape and a geometric solid is that solids look three-dimensional. When we have a graphical representation of a solid, we can calculate important properties like volume, mass, weight, density, and surface area. Let's start by taking a look at each of these properties.

Volume is the amount of space occupied by a substance or object or enclosed within a container.

Volume of a Cube

To find the volume of a cube if all the sides are equal in length, use the following formula. See Figure 2-30.

$$V = s^3$$

V = volume
s = length of side

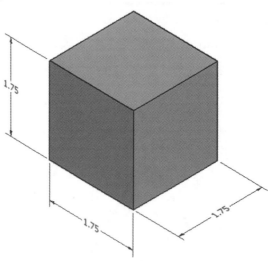

FIGURE 2-30 *Cube.*

V = 1.75 in × 1.75 in × 1.75 in
V = 5.25 in³

Volume of a Rectangular Solid

When calculating the area of rectangular solid, note that one side is longer. Because of this we have to use the following formula. See Figure 2-31.

$$V = wdh$$

v = volume
d = depth
h = height

© 2010 Cengage Learning. All Rights Reserved. May not be scanned, copied or duplicated, or posted to a publicly accessible website, in whole or in part.

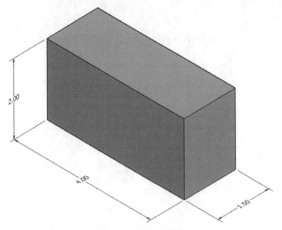

FIGURE 2-31 *Rectangular solid.*

V = wdh
V = 1.5 in × 4 in × 2.0 in
V = 12.0 in³

Volume of a Cylinder

To calculate the volume of a cylinder, use the following formula. See Figure 2-32.

$$V = \pi r^2 h$$

V = volume
π = 3.14
d = diameter
h = height
V = π × diameter × height
V = 3.14 × 3.25 in × 5.5 in
V = 56.12 in³

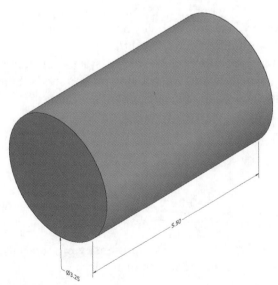

FIGURE 2-32 *Cylinder.*

© 2010 Cengage Learning. All Rights Reserved. May not be scanned, copied or duplicated, or posted to a publicly accessible website, in whole or in part.

Volume of a Cone

To calculate the volume of a cone, use the following formula. See Figure 2-33.

$$V = 1/3\pi r^2 h$$

V = volume
π = 3.14
r = radius
h = height

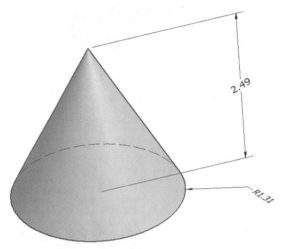

FIGURE 2-33   *Cone.*

$V = (1/3)\ 3.14 \times 1.31^2 \times 2.49$
$V = 4.47\ \text{in}^3$

BACKGROUND

Mass and Weight

People often confuse the terms *mass* and *weight*. Mass is defined as the amount of matter an object contains, and weight is the force of gravity acting on an object. They are both different concepts. For example, a person who weighs 150 pounds on Earth would weigh a whopping 354 pounds on Jupiter! The person has the same mass, but when you factor in gravity on Jupiter their weight will be different. http://www.exploratorium.edu/ronh/weight/.

The units for mass are either grams (metric system) or slugs (English system). The units for weight are Newton's (metric system) or pounds (English system). The following are formulas used to calculate mass and weight:

Weight

$$W = Mg$$

W = Weight
M = mass (in slugs)
g = acceleration due to gravity = 32.16 ft/sec²

If we know the weight density of a solid (density of material), we can use the following formula to calculate the weight of an object:

$$W = VDw$$

W = weight
V = volume
Dw = weight density

© 2010 Cengage Learning. All Rights Reserved. May not be scanned, copied or duplicated, or posted to a publicly accessible website, in whole or in part.

You can find the weight density of a material from a variety of sources including the machinist handbook or Web sites like MatWeb.com. You might need to use an online density conversion tool to convert metric densities to English densities and vice versa.

Convert from	kg/m³	g/cm³	oz/in³	oz/gal (US)	lb/gal (US)	lb/ft³	lb/in³	lb/yd³	ton/yd³
kg/m³	1	0.001	0.000578	0.1335	0.00835	0.0624	0.000036	1.6855	0.00075
g/cm³	1000	1	0.578	133.52	8.35	62.43	0.036	1685.6	0.752
oz/in³	1730	1.73	1	231	14.44	108	0.0625	2916	1.302
oz/gal (US)	7.49	0.00749	0.0043	1	0.0625	0.468	0.00027	12.62	0.0056
lb/gal (US)	119.8	0.12	0.069	16	1	7.48	0.0043	201.97	0.09
lb/ft³	16.02	0.016	0.0093	2.14	0.134	1	0.000579	27	0.0121
lb/in³	27680	27.68	16	3696	231	1728	1	46656	20.82
lb/yd³	0.593	0.00059	0.00034	0.079	0.00495	0.037	0.000021	1	0.00045
ton/yd³	1328.9	1.329	0.768	177.4	11.1	82.96	0.048	2240	1

TABLE 2-1 *Density conversion table.*

To calculate the weight of this soft brass spacing block, use the following formula. See Figure 2-34.

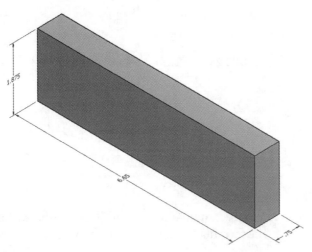

FIGURE 2-34 *Sample problem to calculate weight of block.*

Dimensions:
    L = 6.65
    W = 1.875
    D = .75
Density weight of soft brass is 8,400 Kg/m³.
    W = 9.351 in³ × (8,400 Kg/m³) (.000036 converts to lbs/in³)
    W = 9.351 in³ × .3024 lbs/in³
    W = 2.83 lbs

© 2010 Cengage Learning. All Rights Reserved. May not be scanned, copied or duplicated, or posted to a publicly accessible website, in whole or in part.

Surface Area

Surface area is the measure of the total exposed area of a 3-D solid. For an object with flat faces, such as a cube or rectangle, calculate the area of each surface and then add them all together. To calculate the area of a cylinder or sphere, you need to know the radius.

Surface Area of a Cube

To calculate the surface of a cube, use the following formula. See Figure 2-35.

SA = 6A (six equal sides make up a cube)
SA = surface area
 A = area (length × width) or s^2

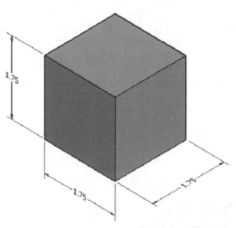

FIGURE 2-35   *Surface area of a cube.*

L W D = 1.75
SA = 6A
SA = 6 × (1.75 in × 1.75 in)
SA = 18.375 in^2

Surface Area of a Rectangular Solid

To calculate the area of a rectangular solid, you need to find the area of each of three different faces. The formula to find area is as follows. See Figure 2-36.

$$SA = 2(wd + wh + dh)$$

SA = surface area
wd = width × depth
wh = width × height
wd = width × height

© 2010 Cengage Learning. All Rights Reserved. May not be scanned, copied or duplicated, or posted to a publicly accessible website, in whole or in part.

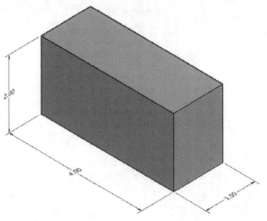

FIGURE 2-36 *Surface area of a rectangular block.*

W = 4.00 cm
H = 2.00 cm
D = 1.50 cm
SA = 2(wd + wh + dh)
SA = 2 × (6 + 8 + 3)
SA = 34 in²

Surface Area of a Cylinder

To find the surface area of a cylinder, you must know the area of the curved face, and the combined area of the circular faces. The formula is as follows. See Figure 2-37.

$$SA = (2\pi r)h + 2(\pi r^2)$$

SA = surface area
r = radius
π = 3.14
h = height

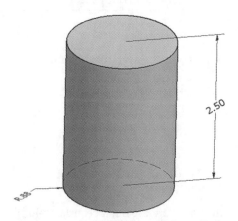

FIGURE 2-37 *Surface area of a cylinder.*

SA = (2πr) h + 2(πr²)
SA = (2)(3.14)(.68)(2.5)in² + 2((3.14)(.68²))in²
SA = 10.676 in² + 8.5408 in²
SA = 19.217 in²

Surface Area of a Sphere

To calculate the surface area of a sphere, you must know the radius and use the following formula. See Figure 2-38.

© 2010 Cengage Learning. All Rights Reserved. May not be scanned, copied or duplicated, or posted to a publicly accessible website, in whole or in part.

$$SA = 4\pi r^2$$

SA = surface area
π = 3.14
r = radius

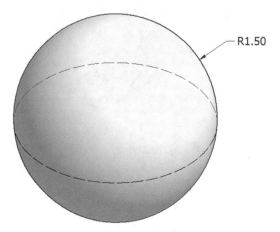

FIGURE 2-38 *Surface area of a sphere.*

Sphere with a diameter of 3 inches, or a radius of 1.5 inches:
SA = $4\pi r^2$
SA = 4 × 3.14 × (1.5 in)²
SA = 4 × 3.14 × 2.25 in²
SA = 28.26 in²

EXERCISE 2.10 CALCULATING VOLUMES, WEIGHTS, AND SURFACE AREAS OF 3-D OBJECTS

Objective

Calculate the needed information through math principles and equations.

Materials

- ☐ Paper
- ☐ Pencil

Procedure

Calculate the needed information for each problem.

1. Cube size is 44 cm. Calculate the surface area, volume, and weight if the cube is aluminum 6013. (Hint: Search Alcoa aluminum for the density and uses of the material.)

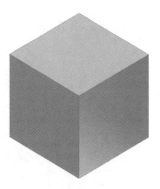

© 2010 Cengage Learning. All Rights Reserved. May not be scanned, copied or duplicated, or posted to a publicly accessible website, in whole or in part.

2. Rectangle size is 1.25 in × 8.25 in × 3.0 in. Calculate the surface area, volume, and weight if the bar is gold. Additionally, calculate the value of the bar using www.kitco.com for the current price.

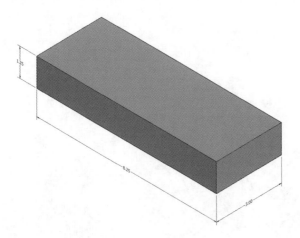

3. Sphere radius is 3.38 in. Calculate the surface area.

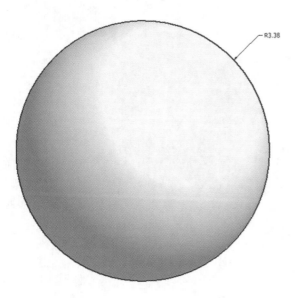

© 2010 Cengage Learning. All Rights Reserved. May not be scanned, copied or duplicated, or posted to a publicly accessible website, in whole or in part.

4. Tube size is 5.60 in. Diameter exterior is 1.60 in. Diameter interior and length is 8.0 inches. Calculate the surface area including the interior surface, the volume, and the weight if the tube is copper. To verify accuracy, create the tube in a 3-D solids package and use the mass properties or analysis tools to verify the hand calculations.

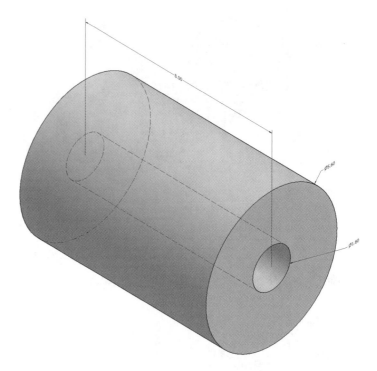

© 2010 Cengage Learning. All Rights Reserved. May not be scanned, copied or duplicated, or posted to a publicly accessible website, in whole or in part.

SECTION 5
Using CAD in Problem Solving

BACKGROUND

Circles and Arcs

From coordinates we move towards circles and arcs. Calculating the centerpoint that the radius occurs from takes geometry and basic math. Try to figure where the centerpoint is on the 4.25 radius arc from the image below. See Figure 2-39.

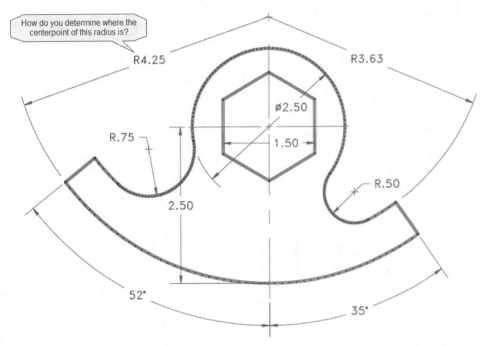

FIGURE 2-39

You need to use some detective work, including studying the images for answers. Remember the children's magazine called *Highlights*? In each issue there was a page that you needed to search to find "hidden" objects. Reading design drawings is very similar. The skills used to find the hidden images are the same ones you are using to determine how to figure out the distance from the centerpoint of the drawing to the centerpoint of the 4.25 radius. See Figure 2-40.

Starting from the center of the hexagon you first work to the bottom arc. That distance is 2.50. From the bottom arc to the centerpoint you find the radius of 4.25. Subtracting 2.50 from 4.25 yields 1.75. This is the distance from the center of the hexagon to the centerpoint of the 4.25-inch line.

© 2010 Cengage Learning. All Rights Reserved. May not be scanned, copied or duplicated, or posted to a publicly accessible website, in whole or in part.

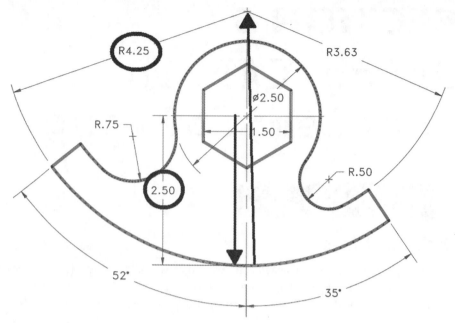

FIGURE 2-40

Using this same example drawing, the 52-degree angle has a line that points from the centerpoint of the 4.25 radius. How do you draw in that line? Math comes to the rescue again, this time bringing angles and subtraction. See Figure 2-41.

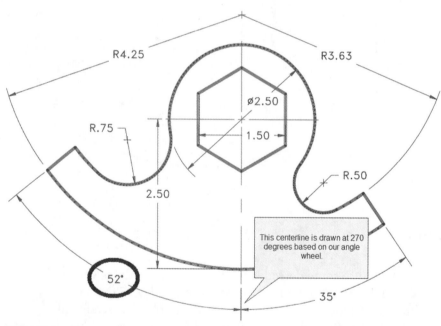

This centerline is drawn at 270 degrees based on our angle wheel.

FIGURE 2-41

From 270, subtract 52, leaving 218 degrees. From the centerpoint, create a line that is 4.25 inches long at 218 degrees and the end of the 4.25 radius arc will be known.

© 2010 Cengage Learning. All Rights Reserved. May not be scanned, copied or duplicated, or posted to a publicly accessible website, in whole or in part.

Using Variables as a Drawing Tool (It Is Similar to Algebra but Easier to Figure Out!)

Anywhere a dimension or a number is input into a parametric modeler a variable or equation could be used to replace a numeric value. First, let's look at equations, and how they can be used to drive dimensions and control a shape. Shown below is the dimensional option dialog obtained by making sure the dimension application is finished (by clicking Done or pressing the Esc key) and then right-clicking on any dimension and clicking on Properties. See Figures 2-42 and 2-43.

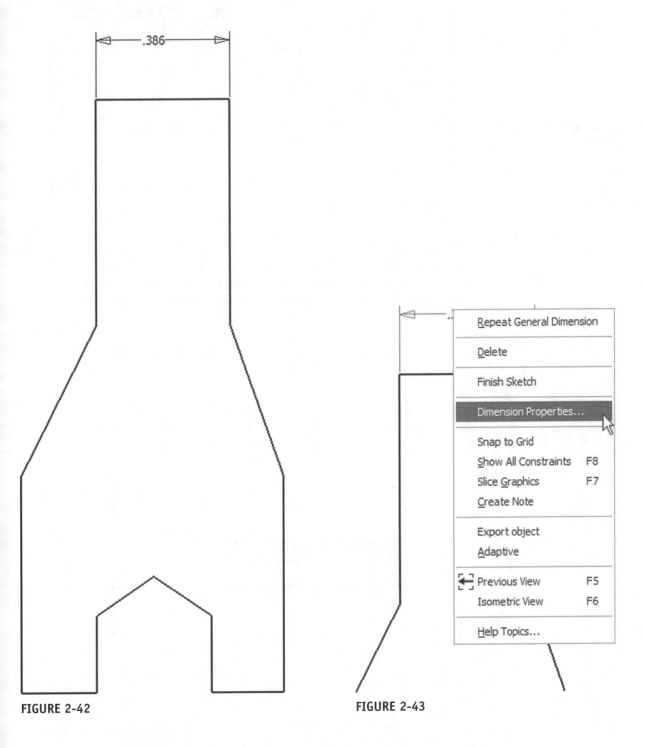

FIGURE 2-42

FIGURE 2-43

© 2010 Cengage Learning. All Rights Reserved. May not be scanned, copied or duplicated, or posted to a publicly accessible website, in whole or in part.

Change the Dimension Properties, under Document Settings to Show Expression. See Figure 2-44. This will reveal the dimensional tag associated with each dimension. Knowing that dimensional tag allows designers to reference and create equations using existing dimensional values.

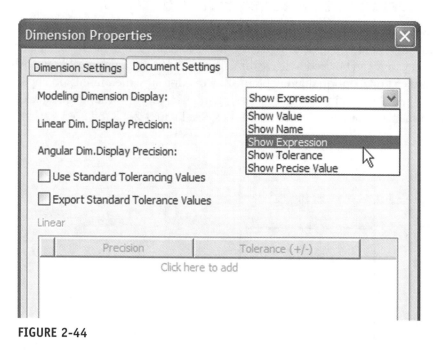

FIGURE 2-44

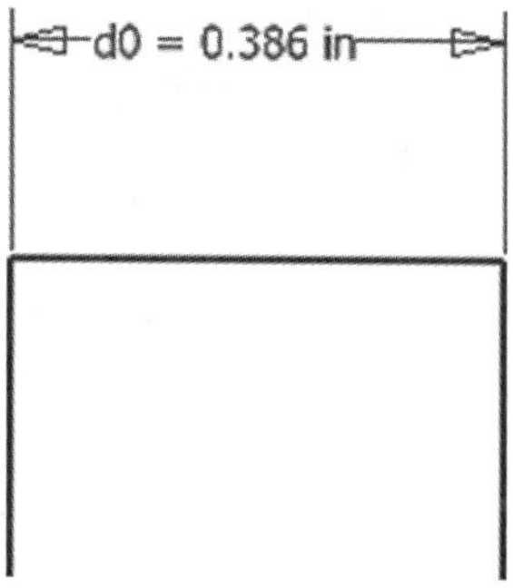

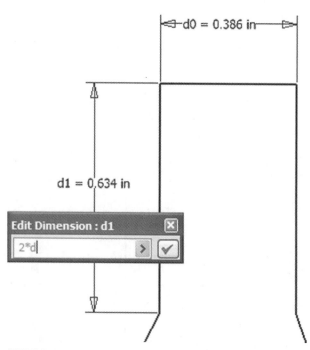

FIGURE 2-45

The first dimension placed is always d0. See Figure 2-45. Let's use that value to drive another dimension through an equation. As you type the equation in it begins red; if you finish typing the equation and it is still colored red, then the equation is incorrectly typed or the variable is wrong. See Figure 2-46.

FIGURE 2-46

© 2010 Cengage Learning. All Rights Reserved. May not be scanned, copied or duplicated, or posted to a publicly accessible website, in whole or in part.

The length of the line will be two times the first dimension of d0. You can make complex equations using multiple dimensional variables, parentheses separating components of the equation. See Figure 2-47.

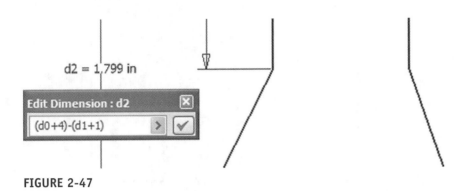

d2 = 1.799 in

Edit Dimension : d2 ☒

(d0+4)-(d1+1)   ⊳ ✓

FIGURE 2-47

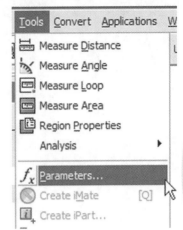

FIGURE 2-48

The second aspect of developing equations is using variables that relate words to a numeric value. In inventor they are called *parameters*. To create a parameter that can be used throughout a design, the command is chosen from the Tools pull-down menu. See Figure 2-48.

Then click the Add button and input a new variable like "wrench". See Figure 2-49.

Parameters ☒

Parameter Name	Unit	Equation	Nominal Value	Tol.	Model Value		Comment
− Model Parameters							
d0	in	0.386 in	0.385711	○	0.385711	☐	
d1	in	2 ul * d0	0.771422	○	0.771422	☐	
d2	in	(d0 + 3 in) - (d1 + 1 in)	1.614289	○	1.614289	☐	
− User Parameters							
▶ wrench						■	

FIGURE 2-49

Select the equation area and then input the numeric value or equation that reflects the size that is applied. See Figure 2-50.

Parameters ☒

Parameter Name	Unit	Equation	Nominal Value	Tol.	Model Value		Comment
− Model Parameters							
d0	in	0.386 in	0.385711	○	0.385711	☐	
d1	in	2 ul * d0	0.771422	○	0.771422	☐	
d2	in	(d0 + 3 in) - (d1 + 1 in)	1.614289	○	1.614289	☐	
− User Parameters							
▶ wrench	in	.50	1.000000	●	1.000000	■	

FIGURE 2-50

© 2010 Cengage Learning. All Rights Reserved. May not be scanned, copied or duplicated, or posted to a publicly accessible website, in whole or in part.

When using these variables just type the name in the Edit Dimension or other location a value is needed. It is really that easy! See Figure 2-51.

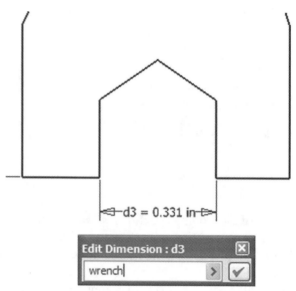

FIGURE 2-51

© 2010 Cengage Learning. All Rights Reserved. May not be scanned, copied or duplicated, or posted to a publicly accessible website, in whole or in part.

SECTION 6
Assembling a Drawing Sheet Packet

BACKGROUND

Creating a sheet packet can be done through either an assembly or from individual part pieces.

The application is the same for both applications. It all surrounds the ability to create base views on new sheets. Creating additional sheets in the same drawing set will create a multiple-sheet drawing set. The steps involved in creating a sheet set are as follows:

STEP 1 Open a drawing sheet and edit the size and configuration of a sheet. (If you have a custom sheet then it can be applied, too.) See Figure 2-52.

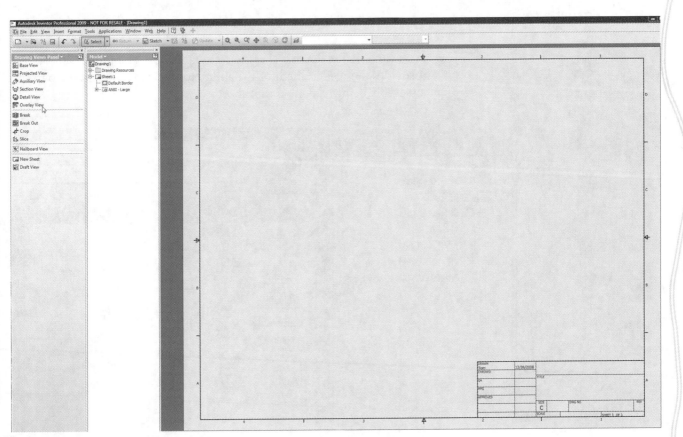

FIGURE 2-52

© 2010 Cengage Learning. All Rights Reserved. May not be scanned, copied or duplicated, or posted to a publicly accessible website, in whole or in part.

STEP 2 Insert a base view on the first sheet. The first sheet should be an assembly view and have a parts list. See Figure 2-53.

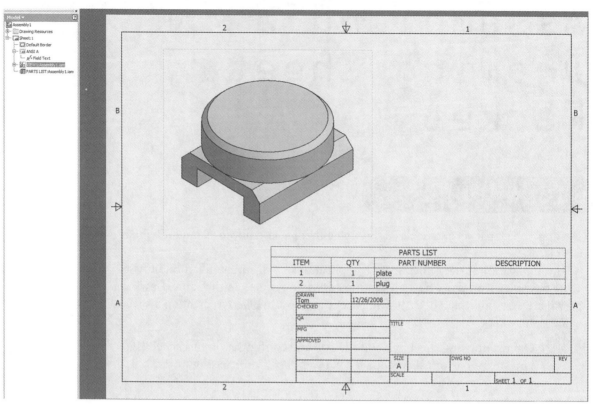

The parts list shown in the figure:

PARTS LIST			
ITEM	QTY	PART NUMBER	DESCRIPTION
1	1	plate	
2	1	plug	

FIGURE 2-53

STEP 3 You can rename the first sheet to Assembly view. Naming the sheet provides the reader or editor the ability to easily understand the part or assembly being shown on that sheet.

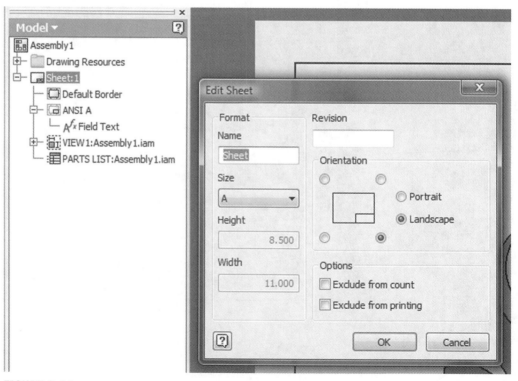

FIGURE 2-54

© 2010 Cengage Learning. All Rights Reserved. May not be scanned, copied or duplicated, or posted to a publicly accessible website, in whole or in part.

STEP 4 Add a new sheet by right-clicking on the topmost line in the browser. See Figure 2-55.

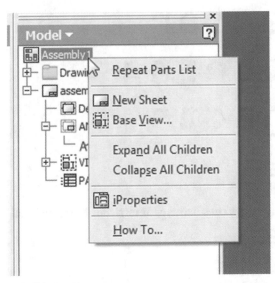

FIGURE 2-55

STEP 5 The new sheet can now have a unique base view and projected view. Don't forget to rename the sheet. Now you have two different sheets in one drawing packet. Add additional sheets, as needed, based on the number of parts or drawings that need to be detailed. The additional drawings will show the number of sheets that have been updated. If you look at the lower right hand corner of Figure 2-56, you will see "SHEET 3 of 3"

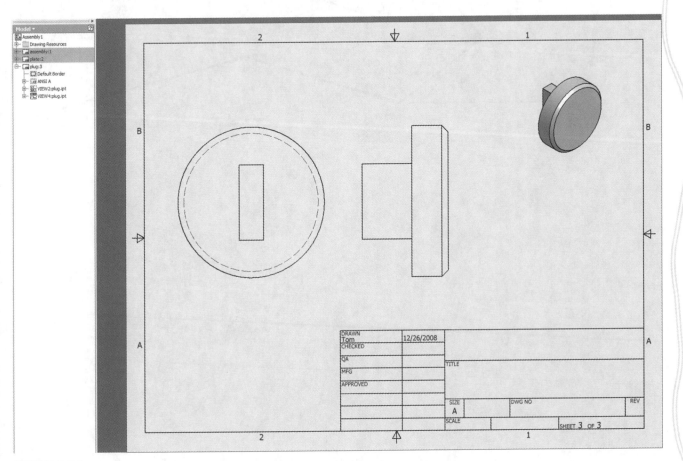

FIGURE 2-56

© 2010 Cengage Learning. All Rights Reserved. May not be scanned, copied or duplicated, or posted to a publicly accessible website, in whole or in part.

SECTION 7
Inventor Sketching Projects

PROCEDURE

Develop constrained sketches using geometric and dimensional constraints for the 2-D shapes.

Problem 2-1

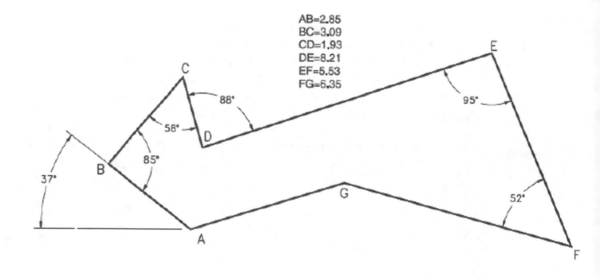

AB=2.85
BC=3.09
CD=1.93
DE=8.21
EF=5.53
FG=6.35

© 2010 Cengage Learning. All Rights Reserved. May not be scanned, copied or duplicated, or posted to a publicly accessible website, in whole or in part.

Problem 2-2

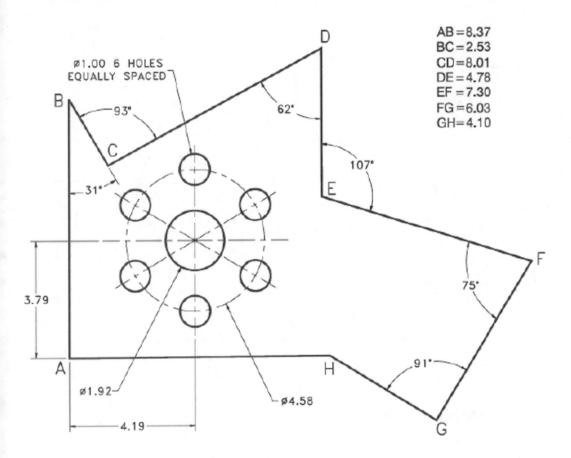

AB = 8.37
BC = 2.53
CD = 8.01
DE = 4.78
EF = 7.30
FG = 6.03
GH = 4.10

Ø1.00 6 HOLES
EQUALLY SPACED

93°
62°
107°
31°
75°
91°
3.79
Ø1.92
Ø4.58
4.19

Problem 2-3

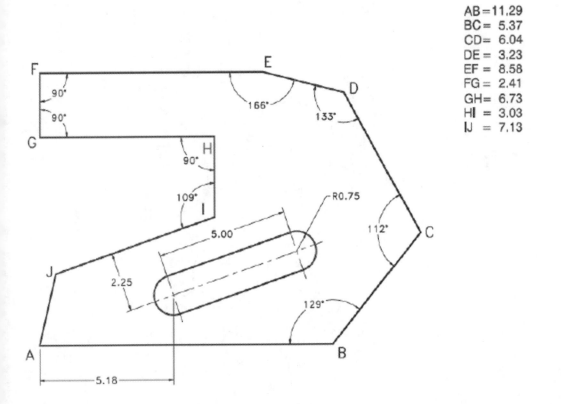

AB = 11.29
BC = 5.37
CD = 6.04
DE = 3.23
EF = 8.58
FG = 2.41
GH = 6.73
HI = 3.03
IJ = 7.13

90°
90°
166°
133°
90°
109°
R0.75
5.00
112°
2.25
129°
5.18

© 2010 Cengage Learning. All Rights Reserved. May not be scanned, copied or duplicated, or posted to a publicly accessible website, in whole or in part.

Problem 2-4

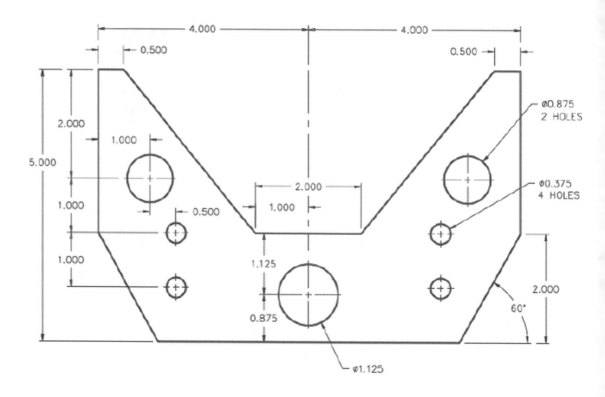

Problem 2-5

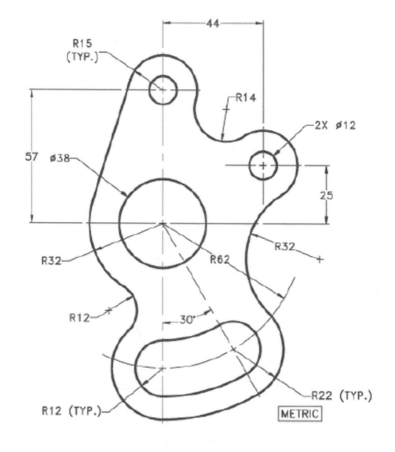

© 2010 Cengage Learning. All Rights Reserved. May not be scanned, copied or duplicated, or posted to a publicly accessible website, in whole or in part.

Problem 2-6

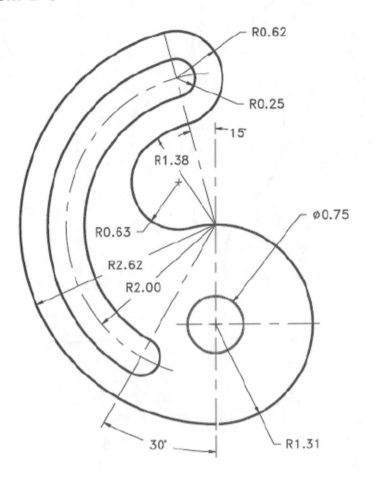

R0.62

R0.25

15°

R1.38

Ø0.75

R0.63

R2.62

R2.00

30°

R1.31

Problem 2-7

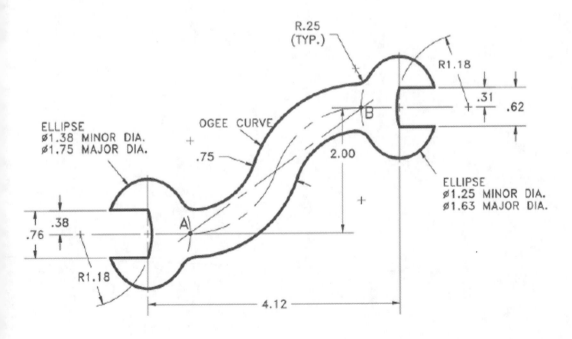

R.25
(TYP.)

R1.18

.31

.62

OGEE CURVE

B

ELLIPSE
Ø1.38 MINOR DIA.
Ø1.75 MAJOR DIA.

.75

2.00

ELLIPSE
Ø1.25 MINOR DIA.
Ø1.63 MAJOR DIA.

.38

.76

A

R1.18

4.12

© 2010 Cengage Learning. All Rights Reserved. May not be scanned, copied or duplicated, or posted to a publicly accessible website, in whole or in part.

Problem 2-8

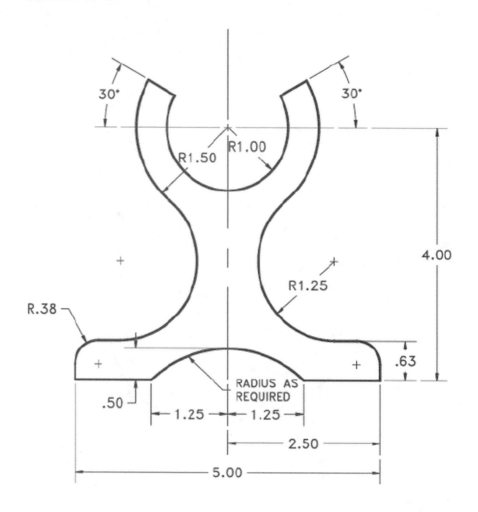

© 2010 Cengage Learning. All Rights Reserved. May not be scanned, copied or duplicated, or posted to a publicly accessible website, in whole or in part.

3-D Part Creation Projects

Problem 2-9

Develop the parts shown as 3-D solid models. As an option, create the multiview drawing as well.

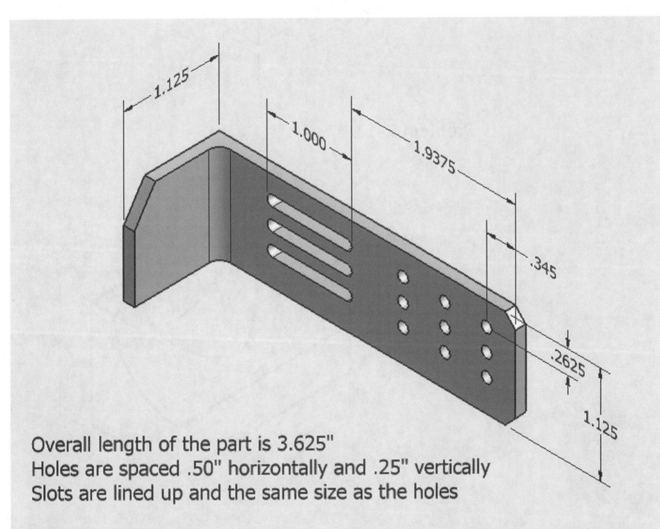

Overall length of the part is 3.625"
Holes are spaced .50" horizontally and .25" vertically
Slots are lined up and the same size as the holes

Fillet is .125" radius
The chamfers near the holes are .125" at 45 degrees
The large chamfer is .25" X .375"

© 2010 Cengage Learning. All Rights Reserved. May not be scanned, copied or duplicated, or posted to a publicly accessible website, in whole or in part.

Problem 2-10

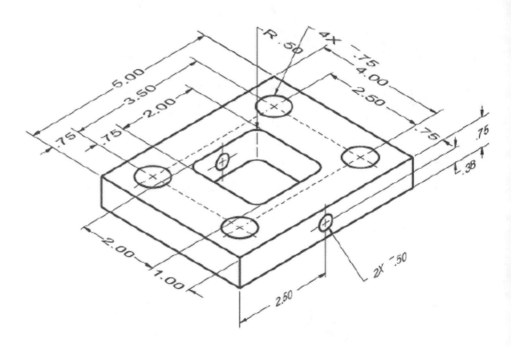

Problem 2-11

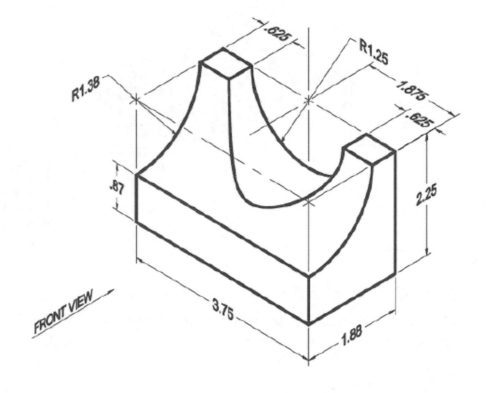

© 2010 Cengage Learning. All Rights Reserved. May not be scanned, copied or duplicated, or posted to a publicly accessible website, in whole or in part.

Problem 2-12

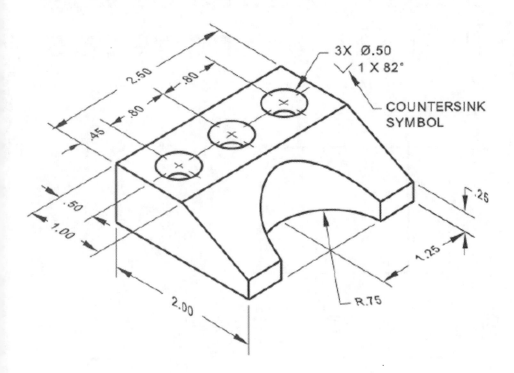

3X Ø.50
⌵ 1 X 82°

COUNTERSINK
SYMBOL

2.50

.80

.80

.45

.50

1.00

2.00

.25

1.25

R.75

© 2010 Cengage Learning. All Rights Reserved. May not be scanned, copied or duplicated, or posted to a publicly accessible website, in whole or in part.

Basic Assembly Drawing Projects

Problem 2-13

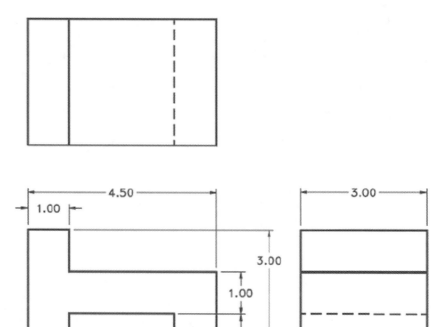

© 2010 Cengage Learning. All Rights Reserved. May not be scanned, copied or duplicated, or posted to a publicly accessible website, in whole or in part.

Problem 2-14

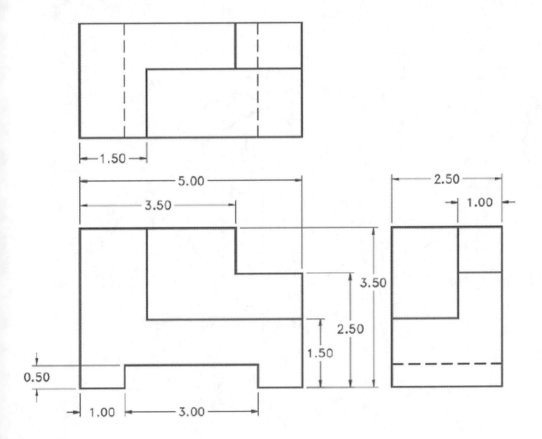

Problem 2-15

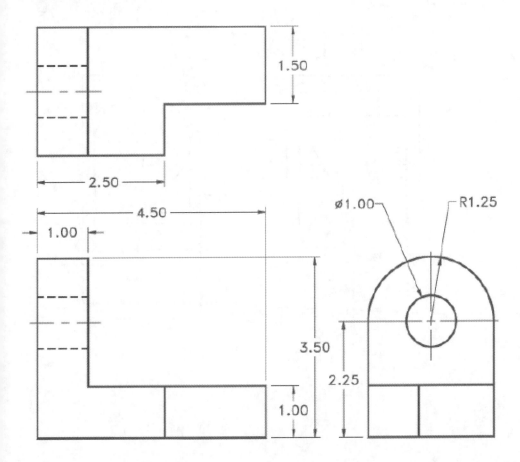

© 2010 Cengage Learning. All Rights Reserved. May not be scanned, copied or duplicated, or posted to a publicly accessible website, in whole or in part.

Problem 2-16

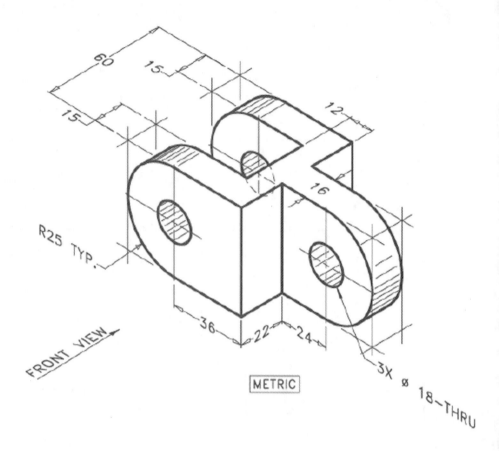

R25 TYP.

60
15
15
12
16
36
22
24

FRONT VIEW

METRIC

3X ø 18-THRU

Problem 2-17

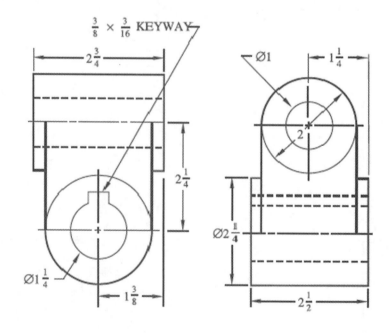

$\frac{3}{8} \times \frac{3}{16}$ KEYWAY

$2\frac{3}{4}$

$2\frac{1}{4}$

$1\frac{3}{8}$

$\varnothing 1\frac{1}{4}$

$\varnothing 1$

$1\frac{1}{4}$

2

$\varnothing 2\frac{1}{4}$

$2\frac{1}{2}$

© 2010 Cengage Learning. All Rights Reserved. May not be scanned, copied or duplicated, or posted to a publicly accessible website, in whole or in part.

Problem 2-18

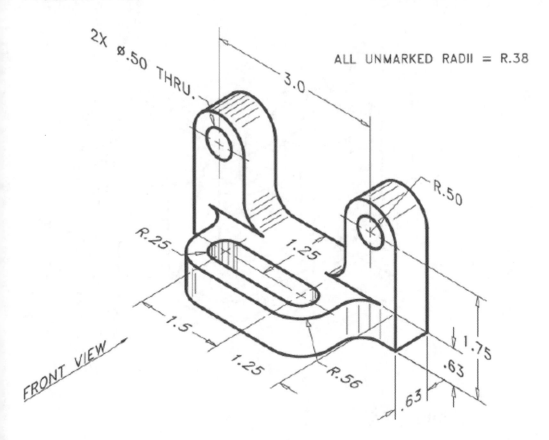

2X ⌀.50 THRU.

ALL UNMARKED RADII = R.38

3.0

R.50

R.25

1.25

1.5

1.25

R.56

1.75

.63

.63

FRONT VIEW

Problem 2-19

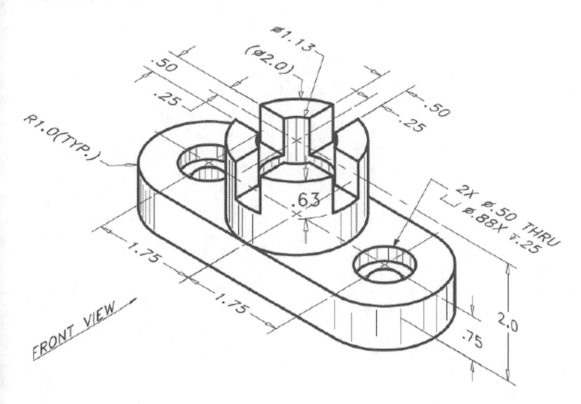

⌀1.13

(⌀2.0)

.50

.25

.50

.25

R1.0(TYP.)

.63

2X ⌀.50 THRU
⌀.88X ▼.25

1.75

1.75

2.0

.75

FRONT VIEW

© 2010 Cengage Learning. All Rights Reserved. May not be scanned, copied or duplicated, or posted to a publicly accessible website, in whole or in part.

Problem 2-20

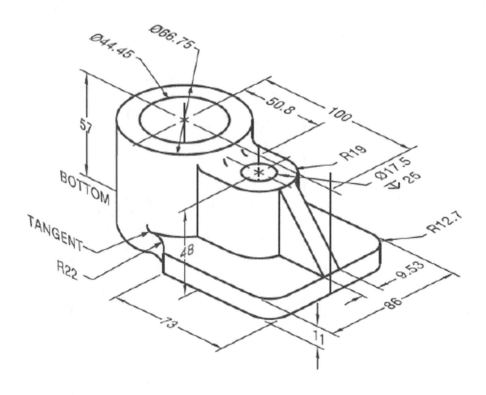

Problem 2-21

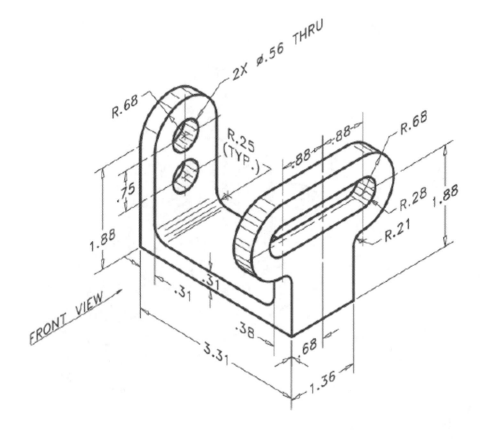

© 2010 Cengage Learning. All Rights Reserved. May not be scanned, copied or duplicated, or posted to a publicly accessible website, in whole or in part.

Problem 2-22

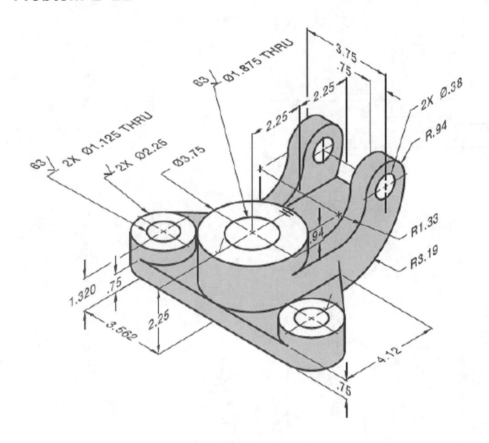

© 2010 Cengage Learning. All Rights Reserved. May not be scanned, copied or duplicated, or posted to a publicly accessible website, in whole or in part.

Problem 2-23

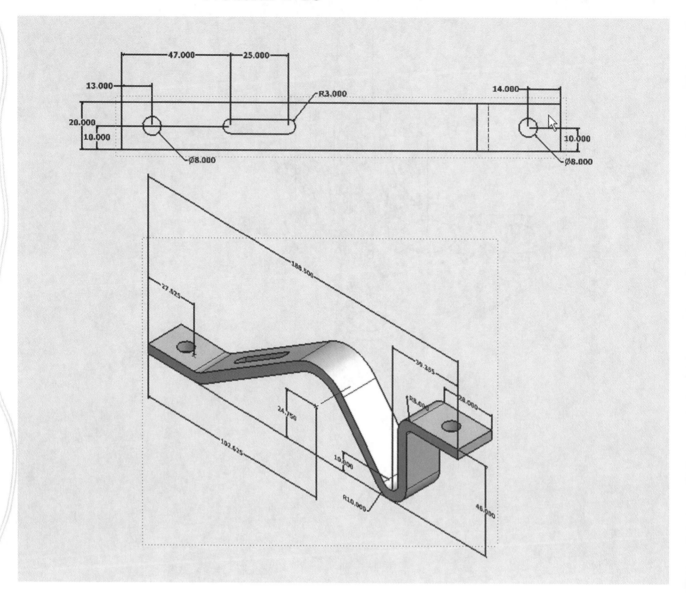

© 2010 Cengage Learning. All Rights Reserved. May not be scanned, copied or duplicated, or posted to a publicly accessible website, in whole or in part.

Problem 2-24

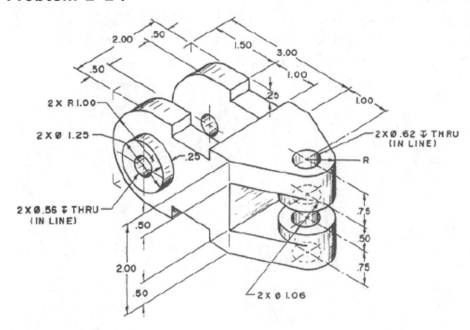

2.00 .50 1.50 3.00 1.00

.50 .25

1.00

2 X R1.00

2 X Ø 1.25

2 X Ø .56 ⊤ THRU
(IN LINE)

.25

R

2 X Ø .62 ⊤ THRU
(IN LINE)

.75

.50

.75

.50

2.00

.50

2 X Ø 1.06

Problem 2-25

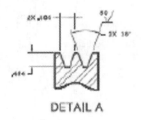

DETAIL A

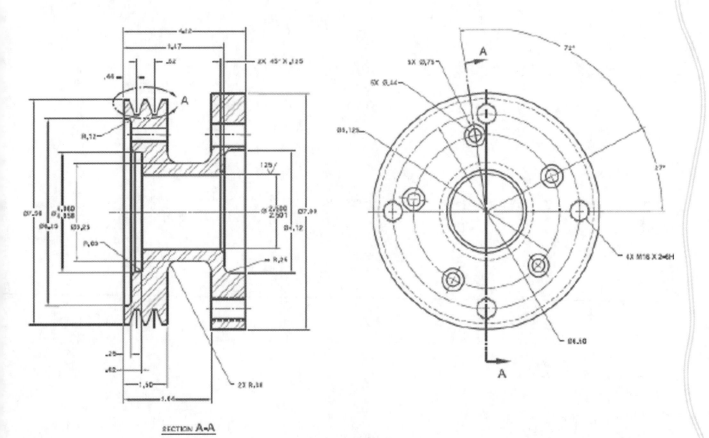

SECTION A-A

© 2010 Cengage Learning. All Rights Reserved. May not be scanned, copied or duplicated, or posted to a publicly accessible website, in whole or in part.

Problem 2-26

Create the golf grip as shown. As an option, create different designs as part of the grip. Emboss your name in the top of the grip. Make sure any embossed items do not affect the thickness of the grip because it could tear.

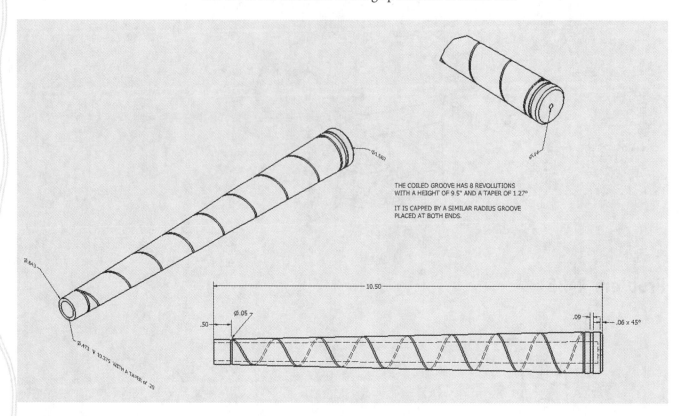

THE COILED GROOVE HAS 8 REVOLUTIONS
WITH A HEIGHT OF 9.5" AND A TAPER OF 1.27°

IT IS CAPPED BY A SIMILAR RADIUS GROOVE
PLACED AT BOTH ENDS.

© 2010 Cengage Learning. All Rights Reserved. May not be scanned, copied or duplicated, or posted to a publicly accessible website, in whole or in part.

Problem 2-27

Here is a drawing of a blade-style putter that was prominent in the 1950s and '60s and into the early '70s. Some golfers still use blade-style putters today. The putters' basic design is simple to use but can increase the error of a person putting if they have a poor swing at the ball. This putter has a thickness of .671 inches in the hosel area (top of the putter blade where it connects to the shaft) and is round with a diameter of .438 inches.

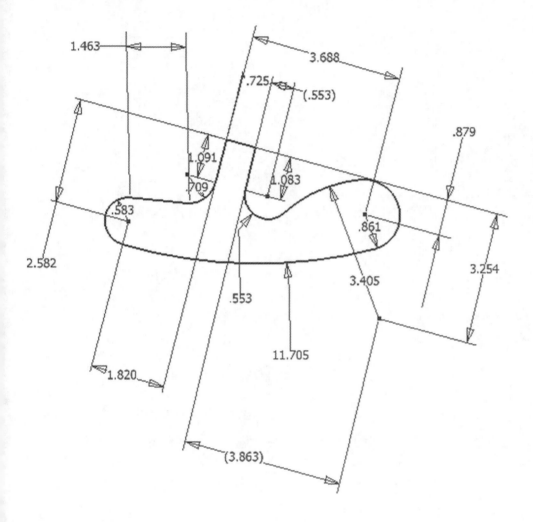

© 2010 Cengage Learning. All Rights Reserved. May not be scanned, copied or duplicated, or posted to a publicly accessible website, in whole or in part.

UNIT 3
Design Solutions Based on Systems Design

Skills List

After completing the activities in this unit, you should be able to:

- Develop basic and intermediate 3D part models

- Create an assembly of multiple parts

- Work with top-down and bottom-up assembly design

- Accomplish a reverse engineering project

- Use aspects of ergonomics and human factors design

- Solve introductory fluids problems

- Identify and apply visual, functional, and structural analysis areas

- Apply basic analysis techniques using design software

© 2010 Cengage Learning. All Rights Reserved. May not be scanned, copied or duplicated, or posted to a publicly accessible website, in whole or in part.

SECTION 1
Ergonomics and Human Factors Design Project

Ergonomics

Ergonomics is the study of how objects interact with people. It is a specialized area of engineering design. Ergonomic studies are done on all types of items that you interact with on a daily basis. Think of all the ergonomic factors that affect cell phone design: Are the buttons too small? Are they spaced comfortably? If it is a flip phone, can an elderly person open it? Can an 8-year-old open it? Does the shape of the phone feel comfortable in your hand? Will the microphone and speaker line up properly alongside an average person's face size?

Ergonomics are also used in factory assembly settings whenever parts are moved or acted on by people. How an item is picked up, moved, or organized is very important in minimizing injuries. Ergonomics researchers study how workers currently perform their tasks and then design improvements to minimize effort and injury. They also train the workers to work safely and efficiently.

EXERCISE 3.1   CELL PHONE ERGONOMICS

Material

☐ Cell phone

Procedure

Be sure to check with your teacher that you can use your cell phone for this exercise. Be sure that the cell phone is *turned off*. Record your results for the following tasks.

Task	Results
1. Pull out your cell phone. Starting from the outer case, identify 2–3 items that you like and 2–4 items you would improve. (For nonopening phone designs do not comment on the buttons or screen interface.)	
2. Then open the phone, do the same task, and comment on the 2–3 good items and 2–4 areas of improvements.	

continued

© 2010 Cengage Learning. All Rights Reserved. May not be scanned, copied or duplicated, or posted to a publicly accessible website, in whole or in part.

Task	Results
3. Now, switch phones with your partner and do the same for your phone, listing both positive and areas of improvements they can identify.	
4. Once finished, combine the areas of improvement and positive comments from the outer and inner reviews. Highlight any similar comments, and in the areas of improvements indicate how you would improve the design compared to the current product.	

EXERCISE 3.2   BOX STACKING

Materials

☐ Scissors

☐ Tape

☐ Page of cutout blocks (See Figure 3-1.)

Procedure

1. On the next page is a series of box patterns. Cut them out carefully and tape them together. You only need one set of blocks for each pair of team members.

2. Number the blocks on one side from 1 to 9.

3. Each team member will perform the following task of stacking the blocks in the correct order from 1 to 9. The team member stacking the blocks cannot bend their arm at the elbow while stacking the blocks, with the numbers facing them in order. Block 9 is on the bottom and block 1 is on the top.

4. The first attempt by each member will be timed. The evaluator will also write down any areas they think could use improvements. Look closely at your team member's form as they perform the task. Record the time.

5. Once both team members have stacked the blocks, discuss the changes that need to be made to improve the time. (For example, pre-aligning the blocks to make it easier to choose faster or changing the body position.)

6. Now try implementing the improvements and start again. Did the time improve? Did you discover any more improvements?

7. The winning team will have the highest level of time improvement combined with the fastest time for stacking the blocks.

© 2010 Cengage Learning. All Rights Reserved. May not be scanned, copied or duplicated, or posted to a publicly accessible website, in whole or in part.

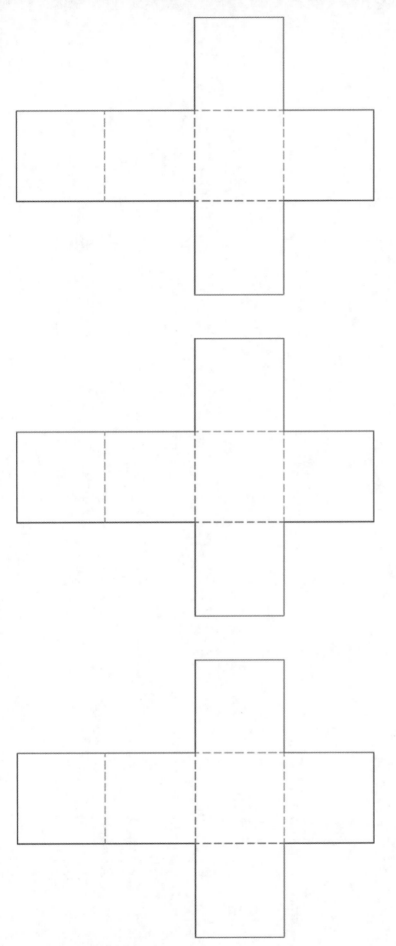

FIGURE 3-1 *Block ergonomics project.*

© 2010 Cengage Learning. All Rights Reserved. May not be scanned, copied or duplicated, or posted to a publicly accessible website, in whole or in part.

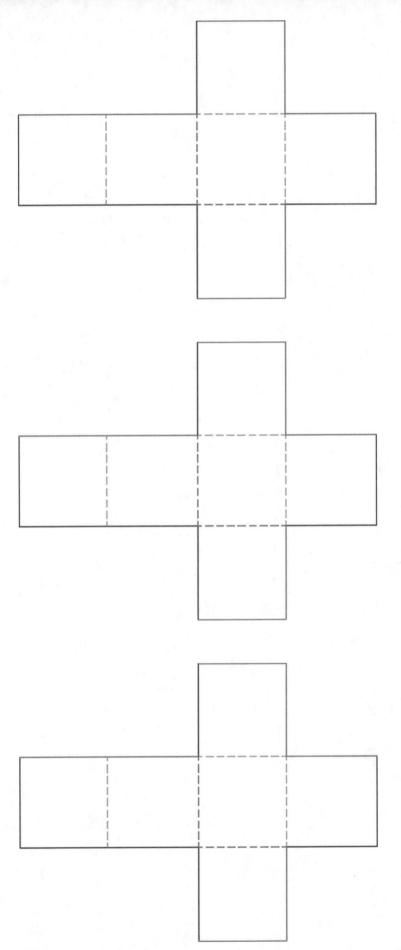

FIGURE 3-1 *continued*

© 2010 Cengage Learning. All Rights Reserved. May not be scanned, copied or duplicated, or posted to a publicly accessible website, in whole or in part.

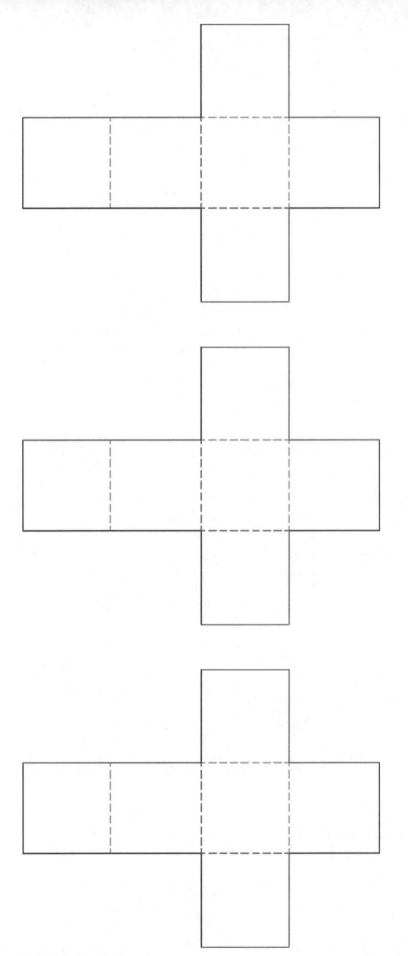

FIGURE 3-1 *continued*

© 2010 Cengage Learning. All Rights Reserved. May not be scanned, copied or duplicated, or posted to a publicly accessible website, in whole or in part.

EXERCISE 3.3   EVERYDAY ERGONOMICS

Materials

- ☐ Plastic spoon
- ☐ Metal spoon
- ☐ Optional spork
- ☐ Sports equipment item (homework)

Procedure

Background: All items that are used by humans have ergonomics involved in the design. You probably can tell the difference between how a plastic fork feels and a metal fork feels when you are eating with it. Why does the metal fork feel better and have better usability?

Your task is to design and investigate ergonomics in several products that are used:

1. Investigate the difference between how a plastic spoon feels and a metal spoon feels, and is used.

   a. Describe the differences.

   b. Indicate the similar characteristics.

   c. What are the positives of the plastic spoon?

   d. What are the positive characteristics of the metal spoon?

   e. How would you improve each of the items ergonomically? (minimum of four items of improvement)

2. To add an additional level of complexity, add a spork into the task of utensils.

3. Complete the task of investigating ergonomics on your favorite piece of sports equipment as homework (baseball bat, golf club, tennis racket, soccer shoe, skateboard, roller blades, etc.).

© 2010 Cengage Learning. All Rights Reserved. May not be scanned, copied or duplicated, or posted to a publicly accessible website, in whole or in part.

SECTION 2
Reverse Engineering

BACKGROUND

Reverse Engineering

The goal of reverse engineering is twofold. One goal is to take one-of-a-kind design items and create production drawings. The other goal is to conduct research on competitive products to enhance your current designs.

EXERCISE 3.4   REVERSE ENGINEER A CELL PHONE

Cell phones come in all different shapes and sizes. Why do you think that is? What consistent design features do all cell phones have?

Materials

☐ Cell phone

☐ Calipers

Procedure

1. IMPORTANT NOTE: In this project, DO NOT open or dismantle any cell phone!

2. Using calipers and rulers: Measure and record, through sketching, the major cell phone external components. This will include the shape and layout of all the items you can see and touch.

3. The deliverable is a set of sketches that has dimensions along with a 2D CAD-based detail sketch with dimensional placement. Creating models is not required.

EXERCISE 3.5   REVERSE ENGINEER SPORTS EQUIPMENT

Materials

☐ Ball sports equipment or similar component

☐ Calipers

☐ Tape measure

☐ Cutting tool (hacksaw or other saw cutting tool)

Whether it is a tennis ball, golf ball, baseball, soccer ball, basketball, badminton birdie, or other sports game component, you probably have at least one of these items at home collecting dust.

© 2010 Cengage Learning. All Rights Reserved. May not be scanned, copied or duplicated, or posted to a publicly accessible website, in whole or in part.

This procedure will require the destruction of the item so make sure there is no sentimental or monetary value attached to this item (don't go cutting up the one-of-a-kind autographed baseball).

Procedure

1. Sketch the entire outside of the object; include the accurate sizes of the item and placement of any manufacturing marks or processes used.

2. Cut the object open. Once cut, sketch the inside. How many layers are there? Is it filled or just open space? Make sure it is sketched and thicknesses are provided.

3. Research if there is any Web information on how the object is manufactured.

4. Create a detailed solid model drawings of the object based on your measurements.

5. Create a written report on how the object is manufactured and assembled. Include your external and internal detail drawings and section drawings if applicable.

EXERCISE 3.6   REVERSE ENGINEER A HOUSEHOLD SPRAY BOTTLE

Materials

- [ ] Clean spray bottle
- [ ] Calipers
- [ ] Tape measure (optional)
- [ ] Small screwdriver
- [ ] Pair of small pliers

Special note: This activity will require the dismantling of a spray mechanism from a household cleaning or beauty product. It will no longer be usable once dismantled. The bottle and spray mechanism should be washed using clean water and emptied so no liquid remains in the bottle or sprayer.

Background

Spray bottle sprayers are devices that we use without thinking about them.

We typically have several in our house or apartment, in a variety of sizes, designs, shapes, and colors.

This project is a pictorial look at reverse engineering a device such as a sprayer.

Pictured is a typical sprayer that can be found in many bottles within a house or apartment. See Figure 3-2.

FIGURE 3-2 *Sprayer removed from bottle.*

© 2010 Cengage Learning. All Rights Reserved. May not be scanned, copied or duplicated, or posted to a publicly accessible website, in whole or in part.

FIGURE 3-3 *Shroud is removed from sprayer, revealing the design.*

As the shroud of the sprayer is removed, be very careful whenever you disassemble a part, especially when it is plastic. You could cause it to warp or crack if too much pressure is applied. Look at the object from all angles to see if there is a locking mechanism or an easy way to remove a part.

Once the shroud is removed, sketch the shroud design (tracing it works too!) and begin to determine if any other parts can be removed from the body without causing any changes in integrity of the parts. See Figure 3-3.

Procedure

This project focuses on creating detailed models of the sprayer mechanism found in all spray bottles. This can be accomplished as a team project consisting of two team members.

STEP 1 The sprayer needs to be dismantled; as the dismantling begins, a record of assembly needs to be made. This indicates how the mechanism is put together. This can be done with hand sketching or digital photographs. See Figures 3-4 through 3-6.

STEP 2 Make sure all parts are inventoried, and stored in a recloseable storage bag.

STEP 3 Once the system is dismantled and stored individually, create sketches and place exact measurements of each part.

STEP 4 Begin building 3D part models of the components.

STEP 5 Create the assembly model of the mechanism.

© 2010 Cengage Learning. All Rights Reserved. May not be scanned, copied or duplicated, or posted to a publicly accessible website, in whole or in part.

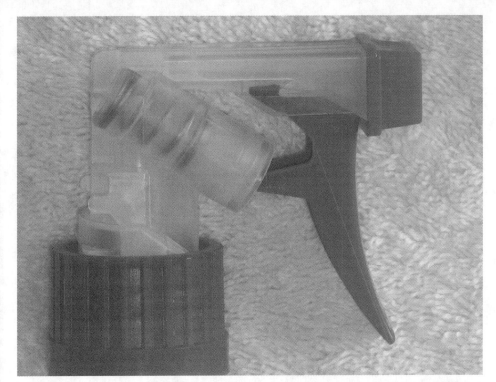

FIGURE 3-4 *Close-up of the sprayer. Notice the spring housing detail and the trigger attachment at the top of the sprayer, with the plunger positioned with the spring.*

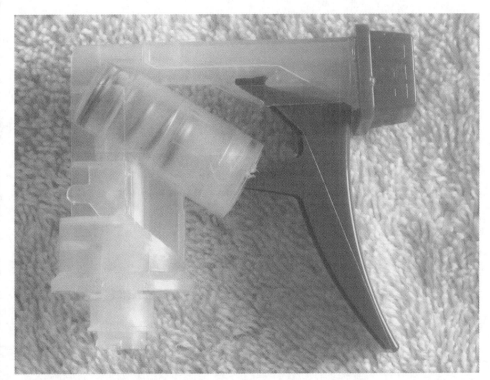

FIGURE 3-5 *Threaded bottle connecter is removed; notice how the sprayer trigger is snapped into place with the barbed ends.*

© 2010 Cengage Learning. All Rights Reserved. May not be scanned, copied or duplicated, or posted to a publicly accessible website, in whole or in part.

FIGURE 3-6 *Top view of the sprayer. This is a thin wall design part.*

© 2010 Cengage Learning. All Rights Reserved. May not be scanned, copied or duplicated, or posted to a publicly accessible website, in whole or in part.

SECTION 3
Analysis
of Models

BACKGROUND

Analysis of the models occurs in two separate areas within design software. First, there is the basic physical properties analysis that is related directly to the material selected (mass). Also found under physical properties is the centroid and principle moments of inertia. The second location is stress analysis of parts under the application's pull-down. These particular tools allow the placement of forces and pressures and fixed points and surfaces to determine the deflection of the object and the areas that show the most stress based on the parameters that are applied.

Mass

Mass is not weight. *Weight* is the gravity effect on a body. *Mass* is the force that keeps an object at constant velocity unless acted upon by an external force. They have the same value but the meaning in the world of engineering is different.

Area

Area is defined as the calculated surface (length × width) of an object. Units are in the fashion of squared units (e.g., square feet or square meters).

Volume

Volume is the displacement of an object (length × width × height). This is measured in cubic units (e.g., cubic feet or cubic meters).

Centroid

The *centroid* can be calculated for 2D or 3D items. It is sometimes called "center of mass" or "center of gravity," which is the point that gravity acts on an object. To understand the meaning of centroid, let's do a hands-on example.

1. Take a map of your state and cut it out.
2. Hold a corner of the cutout and let it swing freely.
3. Draw a line vertically from your finger (use a plumb bob or ruler to help).
4. Then hold the edge of the cutout from another location and let it swing freely.
5. Draw a line vertically from your finger (use a plumb bob or ruler to help).
6. Where the two lines cross is the rough centroid of a 2D body.

Moments of Inertia (MOI)

The explanation is adapted from a presentation made by Jeff Sheets, vice president of research and development at Golfsmith International Inc.

© 2010 Cengage Learning. All Rights Reserved. May not be scanned, copied or duplicated, or posted to a publicly accessible website, in whole or in part.

- Moments of Inertia is the rotational analogue of mass; a measure of the difficulty in rotationally accelerating a body.
- The measurement of MOI is how much mass is distanced from the center of rotation.

Like a skater when their arms are extended, the inertia is spread out and the skater spins slower. When the skater pulls their arms in the body rotates faster. See Figure 3-7.

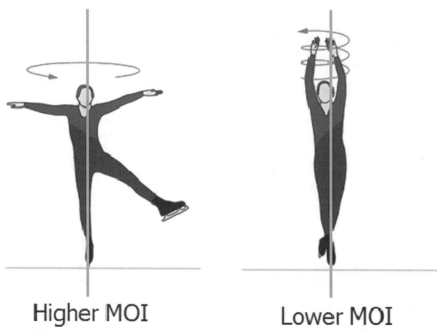

Higher MOI   Lower MOI

FIGURE 3-7 *Examples of moments of inertia.*

A moment of inertia reading is meaningless unless it is defined with respect to a specific rotational axis.

X, Y, Z Coordinates

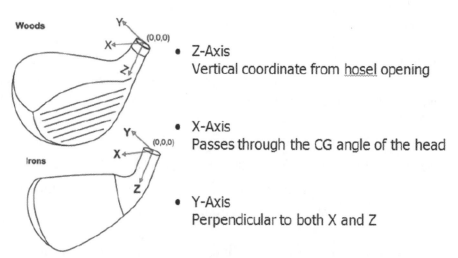

Woods

(0,0,0)

- Z-Axis
  Vertical coordinate from hosel opening

Irons

(0,0,0)

- X-Axis
  Passes through the CG angle of the head

- Y-Axis
  Perpendicular to both X and Z

FIGURE 3-8 *Examples of how axes are applied to golf clubs.*

Golf club heads have the axis positioned so that the Z axis follows the direction of the shaft. The X axis is aligned with the center of gravity of the head. The moments of inertia on a golf club head are applied based on the axes that are aligned to the club head components like hosel and club face. See Figure 3-8.

© 2010 Cengage Learning. All Rights Reserved. May not be scanned, copied or duplicated, or posted to a publicly accessible website, in whole or in part.

EXERCISE 3.7   STRESS ANALYSIS TUTORIAL

Procedure

The basics of using the analysis package inside Inventor will provide some guidance on how improvements can be made to a design to improve the stability and strength of a design. The example product is a design of a hanging plant hook. See Figures 3-9 and 3-10.

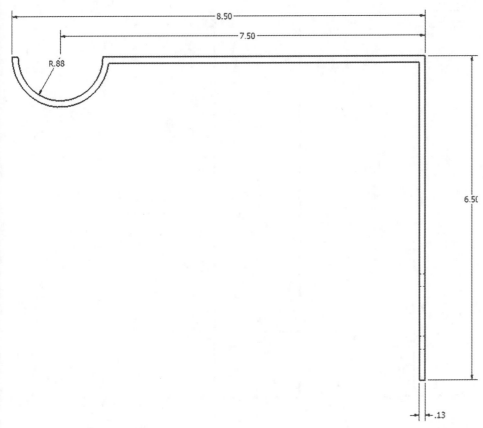

FIGURE 3-9   *Plant hook design.*

After creating the part, move into the stress analysis area from the applications pull-down. Since a material has not been applied to this part design, a dialog appears for material selection. The type of material plays a large part in how a design reacts. If this hook is made from aluminum it would be more flexible (and bend more) than a stainless-steel hook. Other considerations on material selection have to be cost versus the functionality. The plant hook could be made from titanium and thus resist bending, but could another material have a similar response (resist bending) and do it for a lower cost? That is where the analysis tools can assist us in design. It can determine failure scenarios, but it can also help select materials that allow us to produce designs at a lower cost.

© 2010 Cengage Learning. All Rights Reserved. May not be scanned, copied or duplicated, or posted to a publicly accessible website, in whole or in part.

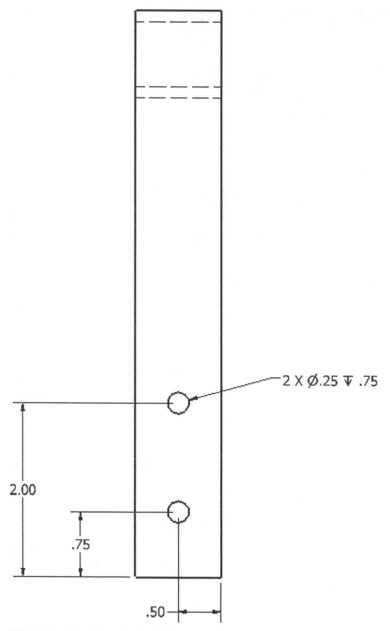

FIGURE 3-10 *Plant hook design.*

For the first trial, select galvanized steel. See Figure 3-11.

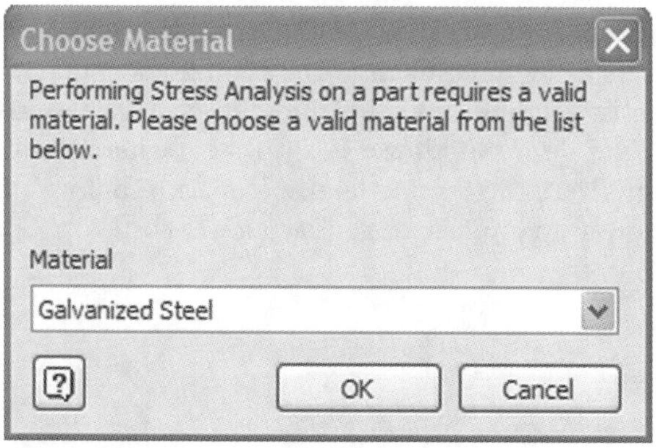

FIGURE 3-11 *Select material for plant hook.*

© 2010 Cengage Learning. All Rights Reserved. May not be scanned, copied or duplicated, or posted to a publicly accessible website, in whole or in part.

The options of the analysis tools are sometimes hard to figure out. In our project we will be using pressure and fixed constraints on our design. See Figure 3-12.

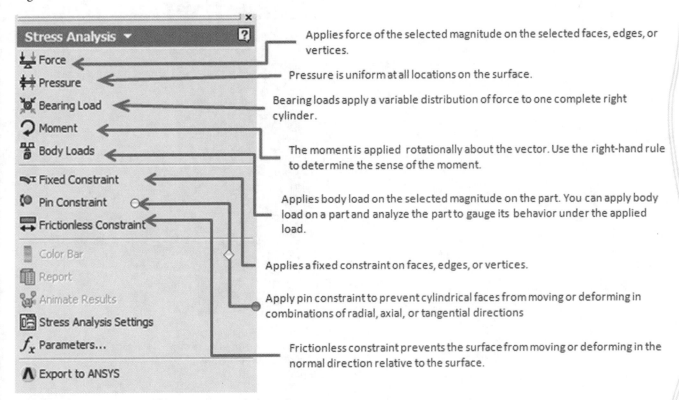

FIGURE 3-12 *Inventor software stress analysis options.*

First, the pressure on the hook will be applied. Figure 3-13. Select the pressure command and locate it approximately in the center of the hook. The pressure of 50 pounds will be our starting force. See Figures 3-14 and 3-15.

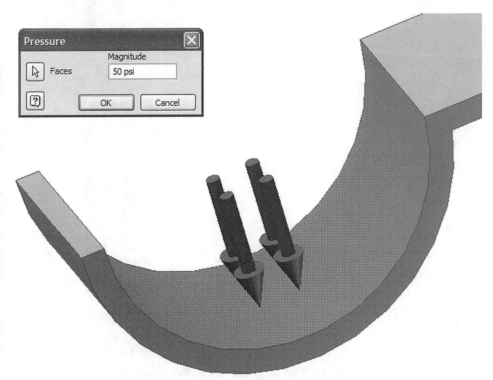

FIGURE 3-13 *Apply pressure to the face of the hook.*

© 2010 Cengage Learning. All Rights Reserved. May not be scanned, copied or duplicated, or posted to a publicly accessible website, in whole or in part.

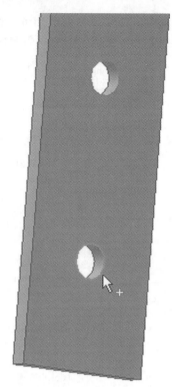

FIGURE 3-14 *Apply fixed constraints on the back edge of the holes on the plant hook.*

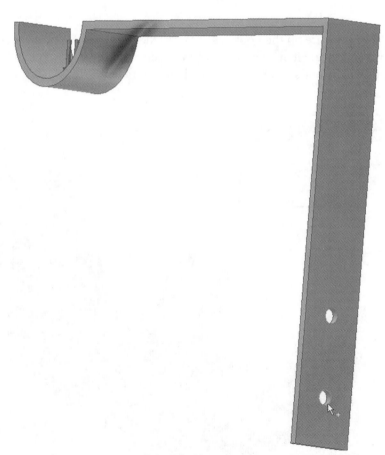

FIGURE 3-15 *Apply fixed constraints on the lower hole of the plant hook.*

© 2010 Cengage Learning. All Rights Reserved. May not be scanned, copied or duplicated, or posted to a publicly accessible website, in whole or in part.

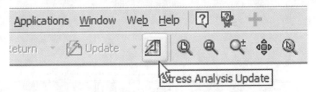

FIGURE 3-16 *Update the analysis after applying the forces and fixed locations.*

Once the fixed constraints are applied, we can now update the stress analysis, which will calculate and show the forces applied on the hook. See Figure 3-16.

When calculations are completed there are several pieces of stress information calculated. In the browser bar the type of information needed is selected. In our first example a warning dialog appeared to alert that the deformation (the distance the hook bends) is very large. To find out what the distance is, select deformation in the browser bar. The calculation indicates that the end of the hook moves over 8 inches. See Figure 3-17.

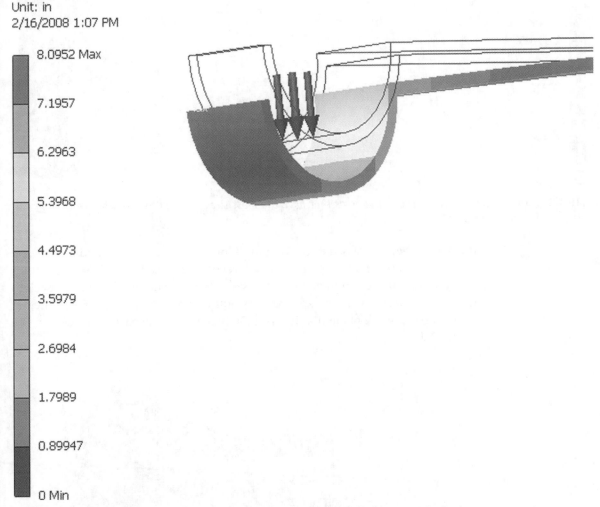

Deformation
Type: Deformation
Unit: in
2/16/2008 1:07 PM

8.0952 Max
7.1957
6.2963
5.3968
4.4973
3.5979
2.6984
1.7989
0.89947
0 Min

FIGURE 3-17 *Deformation of the plant hook with 50 pounds applied is over 8 inches!*

That is unacceptable. That means a 50-pound basket would fall off our hook. The decision is whether we should improve the design or lower the weight limit. Let's improve the design because we need to hold a 50-pound basket. To help determine what needs to get changed, look at the equivalent stress. This shows where

© 2010 Cengage Learning. All Rights Reserved. May not be scanned, copied or duplicated, or posted to a publicly accessible website, in whole or in part.

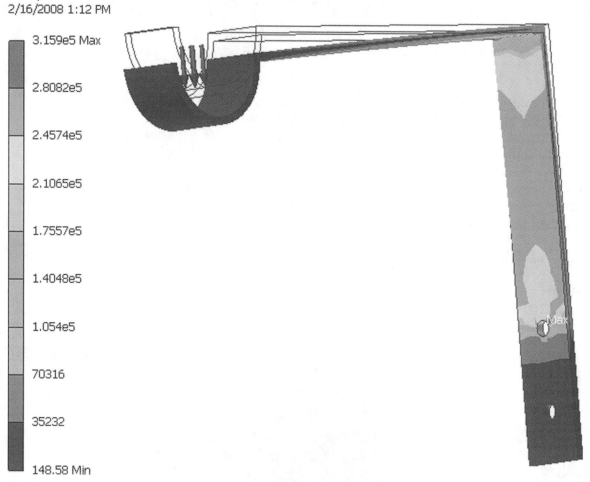

Equivalent Stress
Type: Equivalent Stress
Unit: psi
2/16/2008 1:12 PM

3.159e5 Max

2.8082e5

2.4574e5

2.1065e5

1.7557e5

1.4048e5

1.054e5

70316

35232

148.58 Min

FIGURE 3-18 *Equivalent stress shows the areas that have the most stress applied to them. Notice the highest stress just above the top hole.*

the hook is being stressed in the design. It looks like the top hole and the vertical brace are the areas that have the highest level of stress. See Figure 3-18.

Let's edit our design to enhance the stability of the design. To change back to the editor, select Part in the application's pull-down. We will add a rib to the design. Create a work plane that is offset .50 inches into the part as the center. See Figure 3-19.

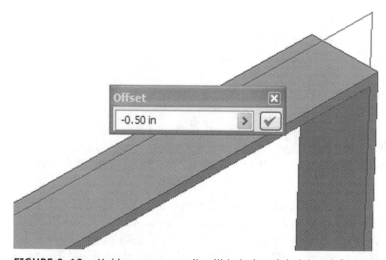

Offset

-0.50 in

FIGURE 3-19 *Making a support rib will help in minimizing deformation.*

© 2010 Cengage Learning. All Rights Reserved. May not be scanned, copied or duplicated, or posted to a publicly accessible website, in whole or in part.

Create a sketch plane on the work plane and draw a rib shape on it. Create a triangular shape with the legs at 3.0 inches each. Use the Rib command with a thickness of .125 inches to generate the rib support. See Figures 3-20 and 3-21.

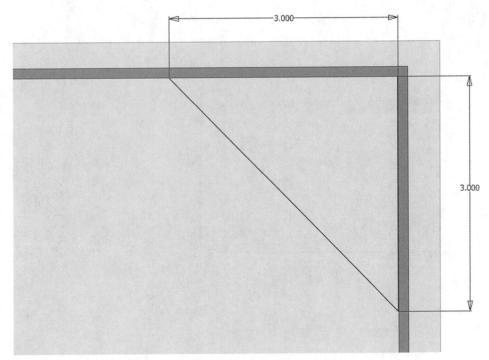

FIGURE 3-20 *Support position on the sketch plane.*

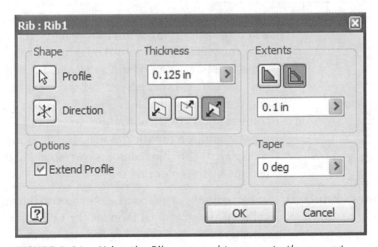

FIGURE 3-21 *Using the Rib command to generate the support brace.*

Update the stress analysis and check the deformation change. The deformation has improved greatly. It is now just over 3 inches, which is still too great for our plant hook. Let's increase the distance of the rib support to 4 inches and the thickness of the rib to .25 inches. Then, update the stress analysis again and see the improvement. See Figure 3-22.

The deformation improved to about 1.5 inches and it is apparent that the brace may need to be very close to the hook to make an impact on the deformation.

From the Inventor help area, there is a good explanation of the charts and stresses that are measured. The safety factor is another good tool to use to help determine whether a part will fail and where it could fail.

© 2010 Cengage Learning. All Rights Reserved. May not be scanned, copied or duplicated, or posted to a publicly accessible website, in whole or in part.

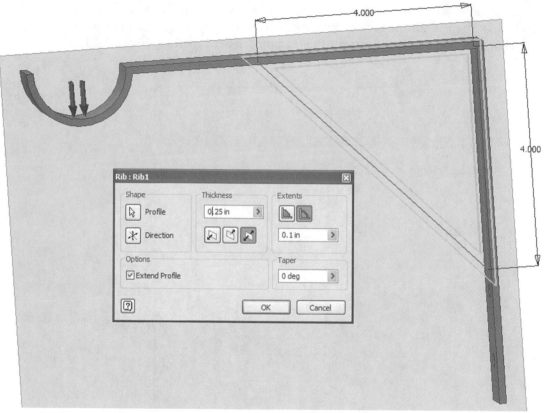

FIGURE 3-22 *Finalizing the rib creation.*

Equivalent Stress	The equivalent stress theory states that failure occurs when the energy of distortion reaches the same energy for yield/failure in uniaxial tension. Equivalent stress can be used to obtain a reasonable estimation of fatigue failure, especially in cases of repeated tensile and tensile-shear loading. Equivalent stress results use color contours to show you the stresses calculated during the solution for your model. The deformed model is displayed. The color contours correspond to the values defined by the color bar.
Maximum Principal Stress	The maximum principal stress gives you the maximum value of the principal stresses. Principal stresses are calculated by transforming the coordinate so that no shear stresses exist. The maximum principal stress gives you the value of stress that is normal to the plane in which the shear stress is zero. The maximum principal stress helps you understand the maximum tensile stress induced in the part due to the loading conditions.
Minimum Principal Stress	The minimum principal stress gives you the minimum value of the principal stresses. The minimum principal stress acts normal to the plane in which shear stress is zero. It helps you understand the maximum compressive stress induced in the part due to the loading conditions.

continued

© 2010 Cengage Learning. All Rights Reserved. May not be scanned, copied or duplicated, or posted to a publicly accessible website, in whole or in part.

Deformation	The deformation results show you the deformed shape of your model after the solution. The color contours show you the magnitude of deformation from the original shape. The color contours correspond to the values defined by the color bar.
Safety Factor	This is the ratio of the yield stress to the maximum equivalent stress. Safety factor shows you the areas of the model that are likely to fail under load. The color contours correspond to the values defined by the color bar.
Frequency Modes	You can view the mode contours for the number of resonant frequencies that you specified in the solution. The modal results appear under the Modes folder in the browser. When you double-click a frequency mode, the mode shape is displayed. The color contours show you the magnitude of deformation from the original shape. The frequency of the mode is shown in the legend. It is also available as a parameter.

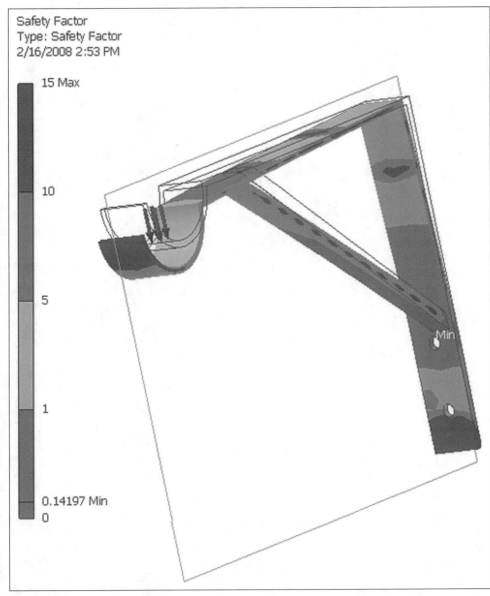

FIGURE 3-23 *Safety factor analysis after the rib is in place.*

© 2010 Cengage Learning. All Rights Reserved. May not be scanned, copied or duplicated, or posted to a publicly accessible website, in whole or in part.

The goal is to minimize the deformation of the plant hook. Design changes could include additional holes located higher up on the bar to affix to a wall or a different brace design. Material choices can also be modified using the Properties tool under the file pull-down.

Any of these changes will adjust the deformation; the key is to work within the constraints specified as part of the design process and always communicate with your customer. Communication will yield additional ideas and ensure that a design meets the customer's needs.

Grading this project can be done based on deformation distance.

© 2010 Cengage Learning. All Rights Reserved. May not be scanned, copied or duplicated, or posted to a publicly accessible website, in whole or in part.

SECTION 4
Fluids

EXERCISE 3.8   FLUIDS: SIMPLE AS A CUP OF JUICE

Materials

- ☐ Measuring cup (large size) per team
- ☐ 3/8-inch ID tubing, length of 3 feet (it can be cut into lengths for the teams)
- ☐ 1 funnel per team with an opening that the tubing fits over
- ☐ Stopwatch or clock with second hand or digital timer

This multipart project begins with the design development of a unique child's cup. It then moves to the second step, which is dispensing the water into the cup based on flow rates.

Procedure

Part 1: Cup Design

The design is for a national family-style restaurant chain. This restaurant has an upscale atmosphere in the dining area that is focused on a combination of seafood and southwestern cuisine.

The task in the project is to come up with a uniquely shaped cup that has a lid and straw. This cup has to hold 6.75 ounces, with an additional space to prevent spilling when a lid is applied.

STEP 1   Research. Collect a variety of children's cups from local eateries. Use the worksheet to gather data.

STEP 2   Plan. Based on your research, what criteria have you decided upon for your cup?

STEP 3   Design. Create four unique designs that hold the required 6.75 ounces of liquid plus your spill-guard distance that you researched. Include a lid and straw design for the cups. The team must have three options for the presentation.

- Be creative. Think outside the box.
- Create detailed drawings with dimensions of all the parts.
- Create an assembly drawing showing the cup, lid, and straw interaction.

STEP 4   Complete the following worksheet.

© 2010 Cengage Learning. All Rights Reserved. May not be scanned, copied or duplicated, or posted to a publicly accessible website, in whole or in part.

1. Identify five items that are the same in all the cups that you collected.

   a.

   b.

   c.

   d.

   e.

2. What unique characteristics does the team's favorite cup have?

3. How many ounces does each cup hold?

4. What is an appropriate spill-prevention distance to apply a lid to a cup?

5. How much space should be allotted below the edge of the cup to prevent spillage?

6. What have you determined about the lid design?

   a. Does the lid fit tightly?

   b. What features must a lid have?

   c. How does the straw interact with the lid?

   d. Does it really prevent spills?

Part 2: The Juice Dispenser Design

This restaurant also needs a dispenser designed for fruit juices that will be placed in the child's cup just designed. The juice comes in 3-gallon plastic bags that have a spout. The spout is 4 inches long and made up of a silicone rubber product so it can be squeezed closed to prevent liquid from dispensing. The exterior diameter of the spout is 5/8 inches. The interior diameter of the spout is 3/8 inches.

© 2010 Cengage Learning. All Rights Reserved. May not be scanned, copied or duplicated, or posted to a publicly accessible website, in whole or in part.

Your task is to figure out a way to dispense 6.75 ounces of liquid into the child's cup from the bag's spout.

Procedure

STEP 1 The design.

    a. Now that the flow rate has been determined for the 3/8-inch tubing, how much time will it take to fill 6.75 ounces?

    b. Your task is to design a mechanism to start and stop the flow of the liquid in the tube for our child's cup to be filled based on the time it was determined to fill the cup.

STEP 2 Deliverables.

    a. Create an assembly model of the mechanism that will dispense the liquid into the cup.

    b. To increase the difficulty of the project, design an enclosure to hold suspend the 3-gallon bag along with the dispensing mechanism.

STEP 3 Use worksheet 2 to help determine some of the research needed for your design.

Worksheet 2: Flow Rates

Determine the flow rate (how many ounces per second) found in the bathroom of your school.

1. To determine the flow rate: use a measuring cup, turn on the water in the sink to a steady flow, place the measuring cup under the flow, and time the flow for 5 or 10 seconds. Make sure the measuring cup does not overflow. Do this test five times and take the average of the rates (ounces per second).

    1.

    2.

    3.

    4.

    5.

2. Do the same test at home to determine the flow rate from several sinks. Flow rate is based on many variables including the age of the pipes in the residence, the elevation of the residence compared to the water tower, and the faucet being tested.

    1.

    2.

    3.

    4.

    5.

© 2010 Cengage Learning. All Rights Reserved. May not be scanned, copied or duplicated, or posted to a publicly accessible website, in whole or in part.

3. How does your home flow rate compare to the school's? What factors do you think influenced the difference?

4. Check the flow rate through the 3/8-inch ID (inside diameter) tubing by performing five tests to determine the average flow rate.

a. This average rate then needs to be compared to the rest of the class's teams to determine the class's average flow rate.

1.

2.

3.

4.

5.

EXERCISE 3.9   HOME ABOVEGROUND POOL PROJECT

Procedure

Create a pool for a family that has a medium-sized backyard. The aboveground pool is elongated; the length is 35 feet, the width is 18 feet, and the corners are 6-foot radius rounded. The pool is 5 feet deep; however, the water depth is 4.5 feet deep. See Figure 3-24.

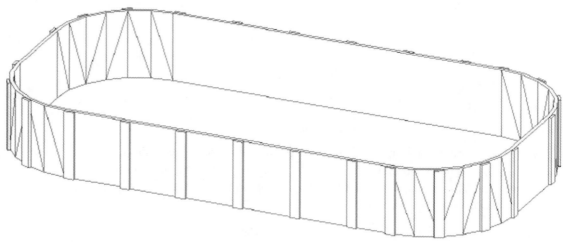

FIGURE 3-24   *Image of sample aboveground pool design.*

Design an aboveground pool shape that meets the size and volume requirements. The vertical stabilizers need to be spaced no further than 4 feet apart on the linear sections and 2 feet apart on the curves.

© 2010 Cengage Learning. All Rights Reserved. May not be scanned, copied or duplicated, or posted to a publicly accessible website, in whole or in part.

Visit the flow rates chart from Akron Brass (flow and reach data) under technical data:
www.akronbrass.com/uploadedFiles/Technical_Data/TD-Theoretical-Discharge-and-Reaction.pdf
This site will help in figuring out answers to the worksheet.

Worksheet: Pool Design

Answer the following questions on your design:

1. How many vertical stabilizers are needed for the pool?

2. How many gallons of water does the pool hold?

3. The theoretical flow you might have at 80 psi is as follows:

   1/2-inch garden hose: 66 gallons per minute

   5/8-inch garden hose: 104 gallons per minute

   3/4-inch garden hose: 149 gallons per minute

4. If you were to fill the pool with your garden hose that is 1/2-inch diameter, how long will it take?

5. How much time is saved if you use a 3/4-inch hose?

6. Instead of using the hose, you are considering bringing in a truck of water. Using the guide above, estimate the gallons per minute for a 4-inch hose. How much time is saved?

7. Draw a detailed sketch of final pool design at the back of book.

© 2010 Cengage Learning. All Rights Reserved. May not be scanned, copied or duplicated, or posted to a publicly accessible website, in whole or in part.

EXERCISE 3.10 HOTEL LAZY RIVER PROJECT

A hotel has contacted you to design a lazy river ride for the pool area. They have an area 250 feet long by 90 feet wide. They would like to have some curves, split paths, etc. in the design. The pool's depth is 3.75 feet. It has to accommodate tubes 4 feet in diameter side by side with some additional space.

Figure 3-25 is a line drawing of the river diagram that the customer shared. You can make improvements to the design. This is a starting sketch.

First, design the lazy river layout, and then calculate the number of gallons of water in your design.

Using a 2.75-inch diameter pipe to fill the river with a flow rate of 120 psi, how long will it take to fill the lazy river?

For flow rate information, visit Akron Brass's data table:

www.akronbrass.com/uploadedFiles/Technical_Data/TD-Theoretical-Discharge-and-Reaction.pdf

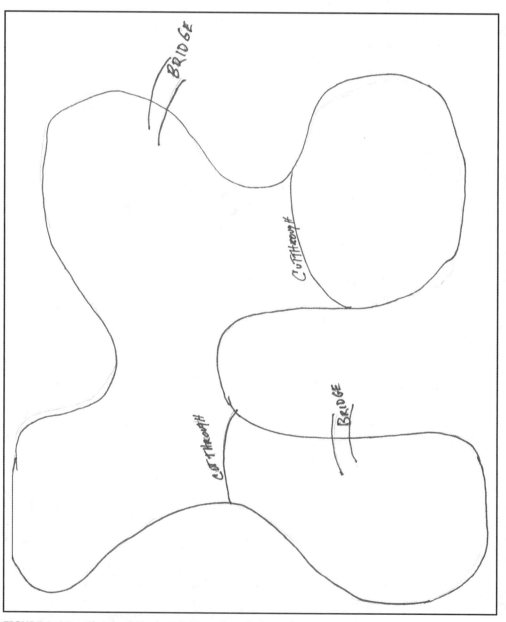

FIGURE 3-25 *Sketch of the hotel's lazy river design.*

© 2010 Cengage Learning. All Rights Reserved. May not be scanned, copied or duplicated, or posted to a publicly accessible website, in whole or in part.

SECTION 5
Working Drawing Projects

PROBLEM 3-1   GOLF PUTTER HEAD

Directions for Problem 3-1

Create a multipiece putter head with removable weighting. General size of the putter is 4.5 inches long, 4.5 inches wide, and about 1 inch high. The inside supports are .25 inches thick and the center T-brace is about 1.25 inches wide and .625 inches tall.

Figure 3-26 is a sketch of a sample design that could be used.

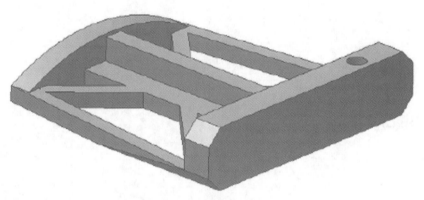

FIGURE 3-26   *A custom designed putter head.*

Critical items to consider:

- Shaft attachment point.
- General weight of the head (the head will be machined aluminum; the designer will determine the best material based on durability, visual, structural, and functional design techniques).
- Provide a total weight of 500 grams for the complete putter head. Calculate the weight of the putter using aluminum.
- Determine where the center of gravity is located.

Figure 3-27 through 3-30 are sketches of the general design.

FIGURE 3-27   *Sketch of the putter's face.*

© 2010 Cengage Learning. All Rights Reserved. May not be scanned, copied or duplicated, or posted to a publicly accessible website, in whole or in part.

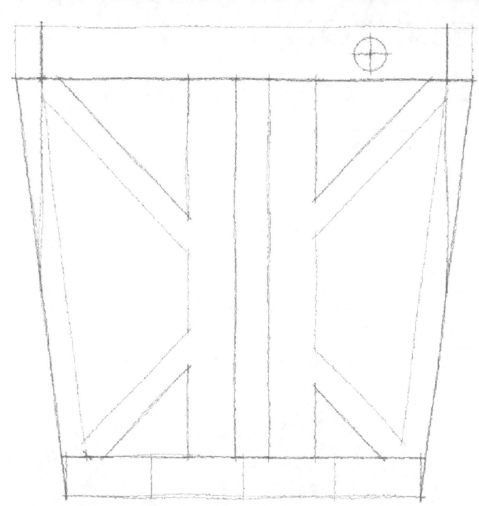

FIGURE 3-28 *Sketch of the top of the putter.*

FIGURE 3-29 *Sketch of the back of the putter.*

FIGURE 3-30 *Sketch of the side of the putter.*

© 2010 Cengage Learning. All Rights Reserved. May not be scanned, copied or duplicated, or posted to a publicly accessible website, in whole or in part.

PROBLEM 3-2   HAMMER

Directions for Problem 3-2

Create an assembly and develop detail drawings for the hammer parts. See Figure 3-31. The handle section has two different styles; develop both styles for the customer for evaluation. Apply the appropriate handle materials or knurl to handles.

Analysis needs to be performed on the handle. Apply pressure at the hammer head end and then at a fixed point at the transition from the handle to the shaft. Determine the force needed that deforms the hammer .25 inches.

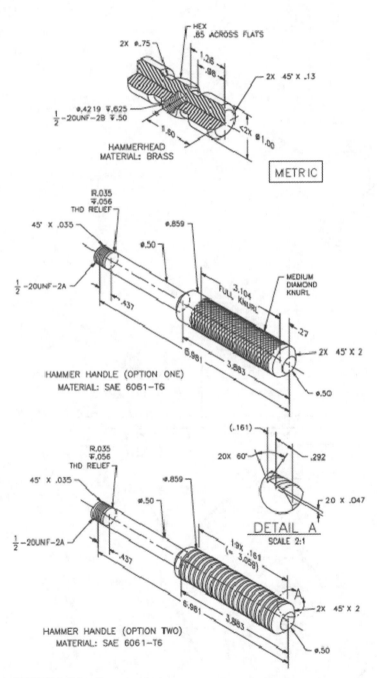

FIGURE 3-31 *Hammer design.*

© 2010 Cengage Learning. All Rights Reserved. May not be scanned, copied or duplicated, or posted to a publicly accessible website, in whole or in part.

Directions for Problem 3-3

Create a solids model and generate an assembly of the caster from the CAD drawing. Additional design information is provided on the CAD drawing for the caster support. See Figure 3-32.

Perform a stress analysis on the wheel supports to see what the maximum weight the caster can hold. Apply pressure to the top of the support and fix the axle hole.

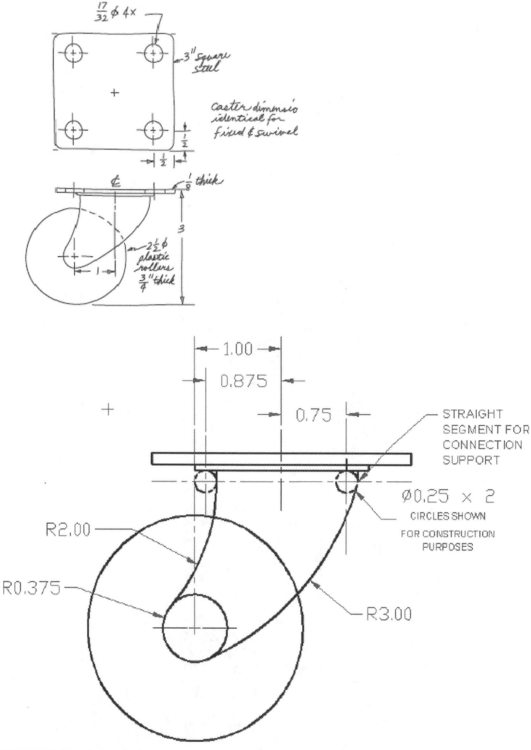

FIGURE 3-32 *Caster design.*

© 2010 Cengage Learning. All Rights Reserved. May not be scanned, copied or duplicated, or posted to a publicly accessible website, in whole or in part.

To increase the complexity of this problem, redesign the caster with a set of bearings between the plate and wheel supports. Experiment with different wheel support designs for the caster. Keep the wheel in the current measured position in any redesign.

PROBLEM 3-4   KEYCHAIN HOLDER

Directions for Problem 3-4

Create a solids model of the keychain assembly. Detail each part for production purposes; create an assembly drawing including a parts list. See Figure 3-33.

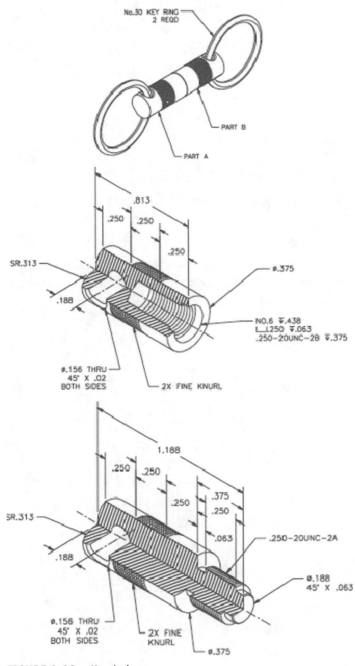

FIGURE 3-33   *Keychain.*

© 2010 Cengage Learning. All Rights Reserved. May not be scanned, copied or duplicated, or posted to a publicly accessible website, in whole or in part.

PROBLEM 3-5   PLUMB BOB

Directions for Problem 3-5

Create the plumb bob using 3D modeling techniques. Create detail drawings of the two objects and an assembly drawing. See Figure 3-34.

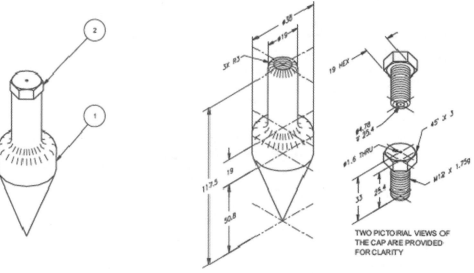

FIGURE 3-34   *Plumb bob.*

PROBLEM 3-6   WOOD CLAMP

Directions for Problem 3-6

Using the sketches, create solids models of each of the pieces of the metric scale clamp assembly. See Figure 3-35. Develop detail drawings of each piece. An assembly model and parts list of all the pieces is also required.

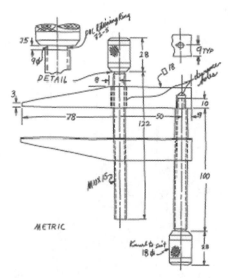

FIGURE 3-35   *Wood clamp.*

© 2010 Cengage Learning. All Rights Reserved. May not be scanned, copied or duplicated, or posted to a publicly accessible website, in whole or in part.

Intermediate-Level Drawings

PROBLEM 3-7 BOAT WINCH

Directions for Problem 3-7

Using the sketches, create parts of each of the pieces of the winch assembly. Develop detail drawings of each piece. See Figure 3-36. Create an assembly and parts list. To add complexity, take the assembly into animation studio and animate the operation of the winch. Creation of the gears can be done through the gear designer.

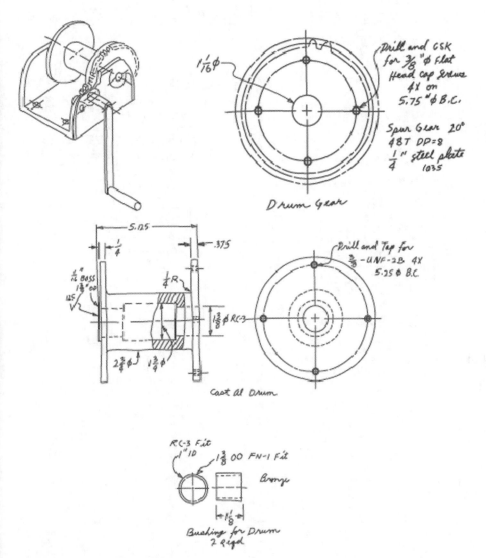

FIGURE 3-36 *Boat winch.*

© 2010 Cengage Learning. All Rights Reserved. May not be scanned, copied or duplicated, or posted to a publicly accessible website, in whole or in part.

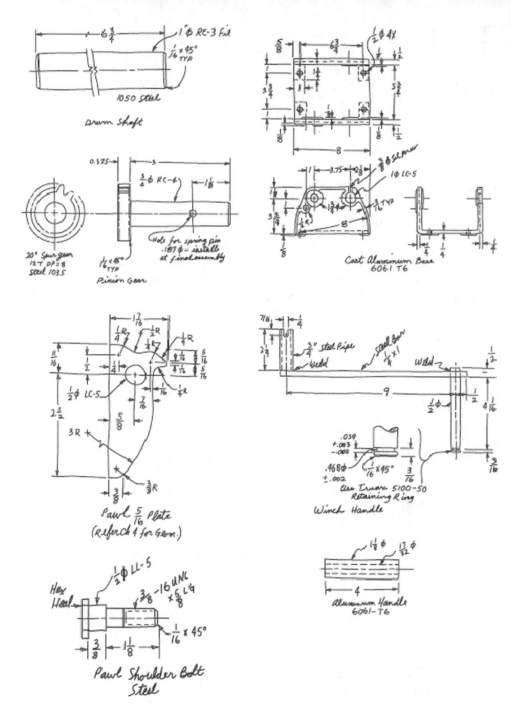

FIGURE 3-36 *continued*

© 2010 Cengage Learning. All Rights Reserved. May not be scanned, copied or duplicated, or posted to a publicly accessible website, in whole or in part.

PROBLEM 3-8 PULLEY

Directions for Problem 3-8

Create an assembly of the roller assembly and create detailed drawings for each of the parts along with an assembly view. See Figure 3-37. The balloons indicate each separate part of the roller assembly.

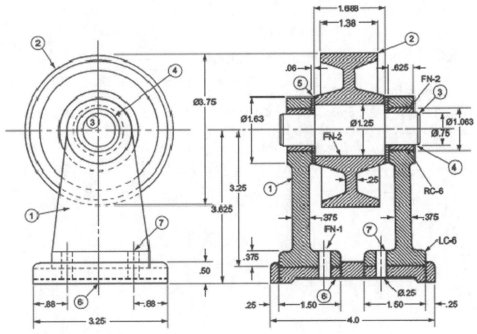

FIGURE 3-37 *Pulley.*

© 2010 Cengage Learning. All Rights Reserved. May not be scanned, copied or duplicated, or posted to a publicly accessible website, in whole or in part.

SECTION 6
Advanced-Level Drawings

PROBLEM 3-9 PEDESTRIAN BRIDGE

Directions for Problem 3-9

Using Figure 3-38, create the model of the pedestrian bridge. If it is created as a single part, it can be tested by using the analysis tools. If it is created as an assembly, the dynamics studio must be used to test the structure.

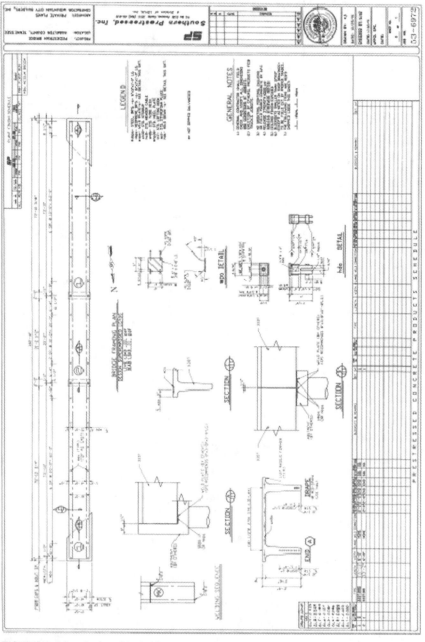

FIGURE 3-38 *Pedestrian bridge.*

© 2010 Cengage Learning. All Rights Reserved. May not be scanned, copied or duplicated, or posted to a publicly accessible website, in whole or in part.

PROBLEM 3-10 GAGE

Directions for Problem 3-10

Create the Mill Work Stop assembly as a series of solids model parts. See Figure 3-40. The knobs are purchase parts, so their dimension will come from specifications sheets. These specifications sheets are available through parts suppliers like Carr Lane Manufacturing through the Internet. See Figure 3-39.

ITEM	QTY	NAME	DESCRIPTION	MATERIAL	NOTES
1	1	VERTICAL MEMBER	3/4 × 2 × 4	SAE 6061	
2	1	BASE MEMBER	1 × 2 × 3.5	SAE 6061	
3	2	RIB	1/4 PLATE	SAE 6061	
4	1	ARM	3/8 × 1 × 5–1/2	SAE 1018	
5	1	CLAMP	1/2 × 1 × 1–5/16	SAE 1018	
6	1	TEE PLATE	5/16 × 1 × 1–1/4	SAE 1018	DRILL AND TAP AT CENTER FOR 5/16–18UNC-2 THREAD
7	1	STOP ROD	$\varnothing$5/16 × 6	SAE 303 SS	
8	1	WING NUT	1/4 − 28 × 1/2		
9	1	BUSHING	$\varnothing$1/2 × .31	SAE 1018	
10	1	TEE STUD	$\varnothing$3/8−16 × 2 LG ALL THREAD		
11	1	CARRIAGE BOLT	5/16−18UNC-2		
12	1	HAND KNOB	HK-4A	ALUMINUM	3/8–16UNC-2B
13	1	HAND KNOB	HK-2A	ALUMINUM	5/16–18UNC-2B

FIGURE 3-39 *Problem 3-10 parts list.*

© 2010 Cengage Learning. All Rights Reserved. May not be scanned, copied or duplicated, or posted to a publicly accessible website, in whole or in part.

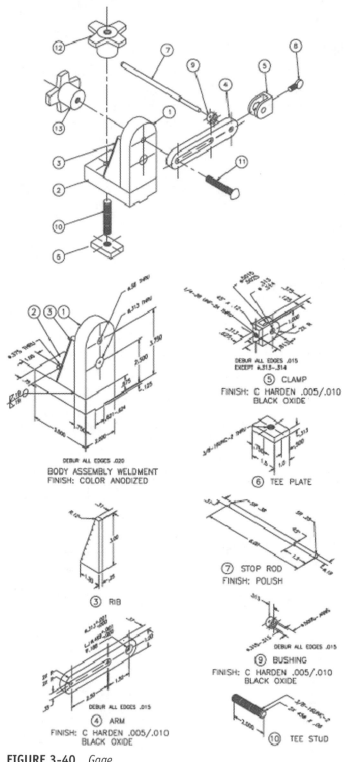

FIGURE 3-40 *Gage.*

PROBLEM 3-11 VALVE

Directions for Problem 3-11

Create an assembly that makes up the ball valve. Each part will have a detailed drawing. Use the dimensioned drawing and the parts list for sizing information. See Figure 3-41.

© 2010 Cengage Learning. All Rights Reserved. May not be scanned, copied or duplicated, or posted to a publicly accessible website, in whole or in part.

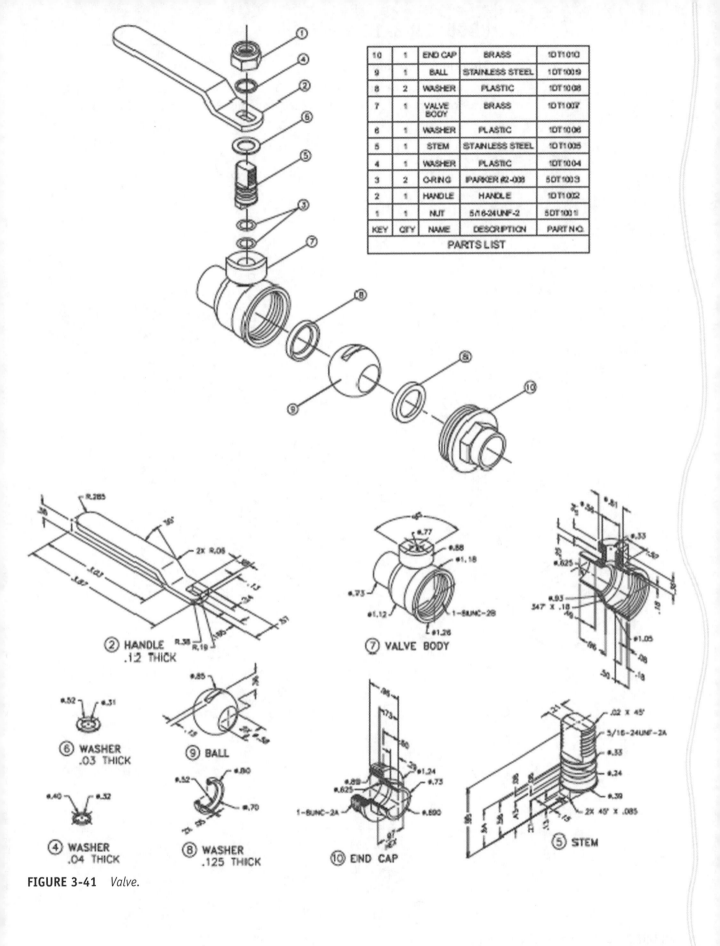

10	1	END CAP	BRASS	1DT1010
9	1	BALL	STAINLESS STEEL	1DT1009
8	2	WASHER	PLASTIC	1DT1008
7	1	VALVE BODY	BRASS	1DT1007
6	1	WASHER	PLASTIC	1DT1006
5	1	STEM	STAINLESS STEEL	1DT1005
4	1	WASHER	PLASTIC	1DT1004
3	2	O-RING	PARKER #2-008	5DT1003
2	1	HANDLE	HANDLE	1DT1002
1	1	NUT	5/16-24UNF-2	5DT1001
KEY	QTY	NAME	DESCRIPTION	PART NO.
			PARTS LIST	

② HANDLE
.12 THICK

⑦ VALVE BODY

⑥ WASHER
.03 THICK

⑨ BALL

④ WASHER
.04 THICK

⑧ WASHER
.125 THICK

⑩ END CAP

⑤ STEM

FIGURE 3-41 *Valve.*

© 2010 Cengage Learning. All Rights Reserved. May not be scanned, copied or duplicated, or posted to a publicly accessible website, in whole or in part.

Directions for Problem 3-12

Create each of the parts of the worm gear reducer as an assembly model. Select three or more parts to create detailed drawings. See Figures 3-42 and 3-43.

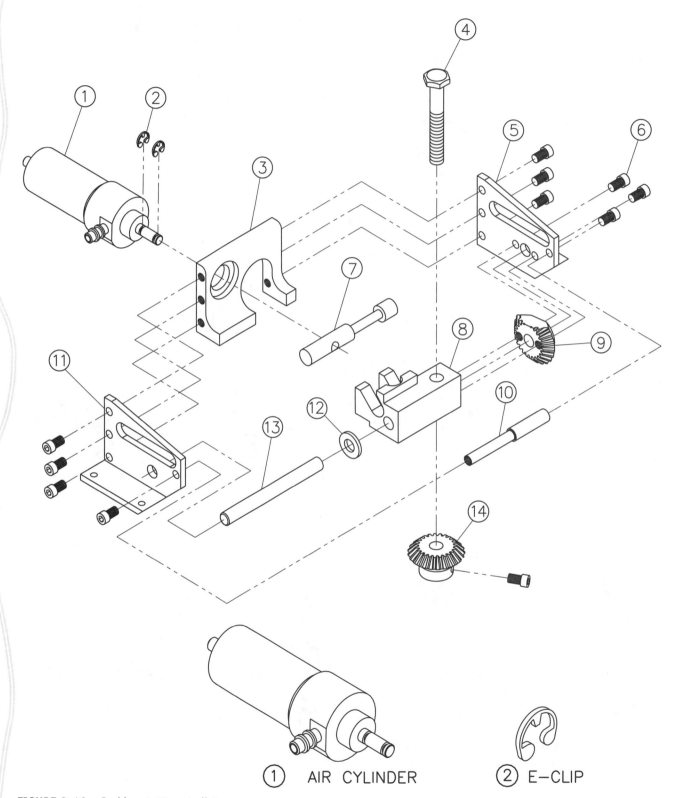

FIGURE 3-42 *Problem 3-12 parts list.*

© 2010 Cengage Learning. All Rights Reserved. May not be scanned, copied or duplicated, or posted to a publicly accessible website, in whole or in part.

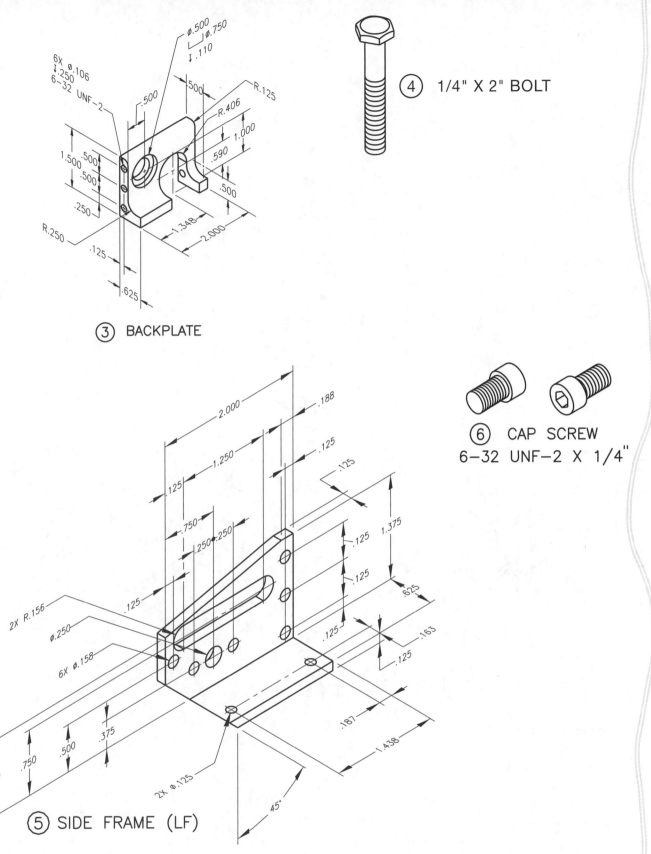

(4) 1/4" X 2" BOLT

(3) BACKPLATE

(6) CAP SCREW
6-32 UNF-2 X 1/4"

(5) SIDE FRAME (LF)

FIGURE 3-42 *continued*

© 2010 Cengage Learning. All Rights Reserved. May not be scanned, copied or duplicated, or posted to a publicly accessible website, in whole or in part.

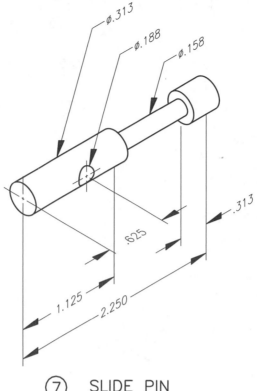

⑦ SLIDE PIN

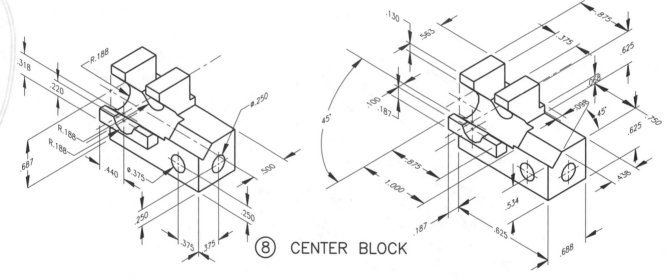

⑧ CENTER BLOCK

FIGURE 3-42 *continued*

© 2010 Cengage Learning. All Rights Reserved. May not be scanned, copied or duplicated, or posted to a publicly accessible website, in whole or in part.

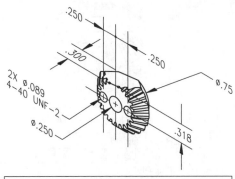

.250
.300
.250
2X Ø.089
4-40 UNF-2
Ø.250
Ø.75
.318

BEVEL GEAR DATA	
NUMBER OF TEETH	20
DIAMETRAL PITCH	20
PRESSURE ANGLE	20d
CONE DISTANCE	.707
PITCH DIAMETER	1.000
CIRCULAR THICKNESS (REF)	.07854
PITCH ANGLE	45d
ROOT ANGLE	40d
ADDENDEM	.05
WHOLE DEPTH	2.188
CHORDAL ADDENDEM	.052
CHORDAL THICKNESS	.078
DEDENDUM	.0625
OUTSIDE DIAMETER	1.071

⑨ 45° BEVEL GEAR (MODIFIED)

.250
R.010
Ø.188
4-40 UNF-2
Ø.089
↧.250
.875
2.000

⑩ SPACER

⑫ WASHER

1/4" BRASS

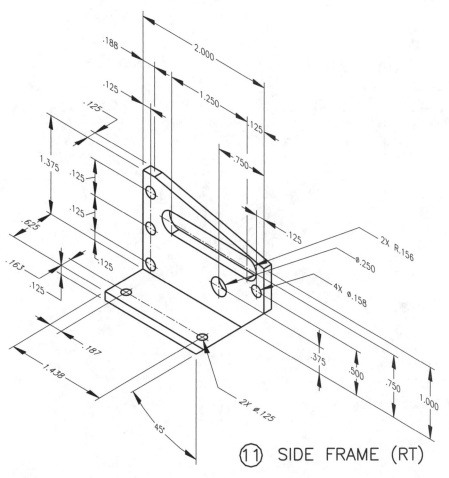

.188
2.000
.125
1.250
.125
.125
.750
1.375
.125
.125
.625
.125
.163
.125
.187
1.438
45°
2X Ø.125
2X R.156
Ø.250
4X Ø.158
.375
.500
.750
1.000

⑪ SIDE FRAME (RT)

FIGURE 3-42 *continued*

© 2010 Cengage Learning. All Rights Reserved. May not be scanned, copied or duplicated, or posted to a publicly accessible website, in whole or in part.

BEVEL GEAR DATA	
NUMBER OF TEETH	20
DIAMETRAL PITCH	20
PRESSURE ANGLE	20d
CONE DISTANCE	.707
PITCH DIAMETER	1.000
CIRCULAR THICKNESS (REF)	.07854
PITCH ANGLE	45d
ROOT ANGLE	40d
ADDENDEM	.05
WHOLE DEPTH	2.188
CHORDAL ADDENDEM	.052
CHORDAL THICKNESS	.078
DEDENDUM	.0625
OUTSIDE DIAMETER	1.071

Ø.250

2.000

45° X .032

(13) PIVOT PIN

(14) 45° BEVEL GEAR

FIGURE 3-42 *continued*

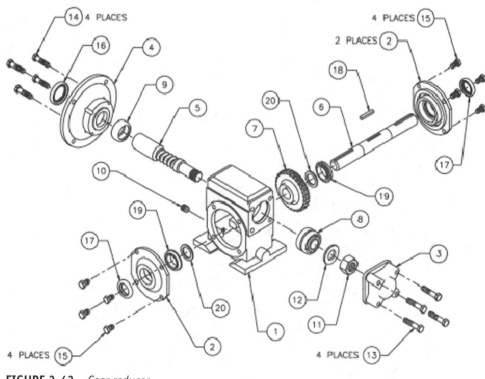

FIGURE 3-43 *Gear reducer.*

© 2010 Cengage Learning. All Rights Reserved. May not be scanned, copied or duplicated, or posted to a publicly accessible website, in whole or in part.

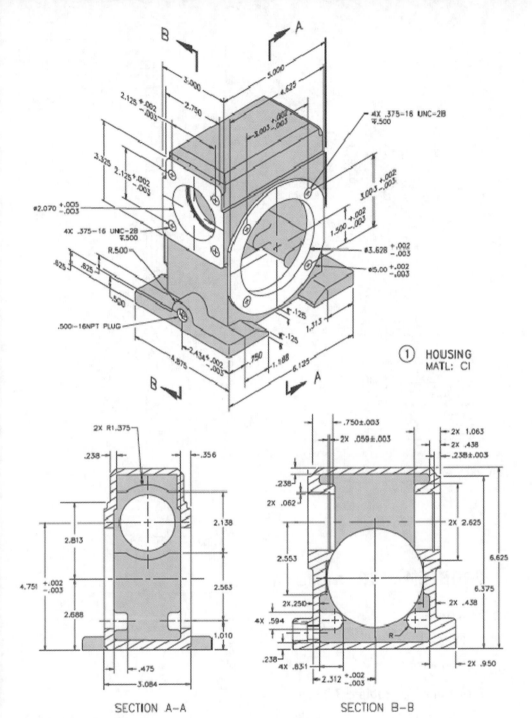

FIGURE 3-43 *continued*

SECTION A-A

SECTION B-B

① HOUSING
MATL: CI

© 2010 Cengage Learning. All Rights Reserved. May not be scanned, copied or duplicated, or posted to a publicly accessible website, in whole or in part.

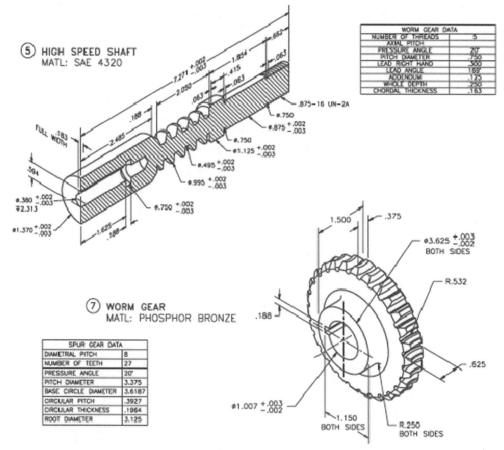

⑤ HIGH SPEED SHAFT
MATL: SAE 4320

WORM GEAR DATA	
NUMBER OF THREADS	5
AXIAL PITCH	
PRESSURE ANGLE	20°
PITCH DIAMETER	.750
LEAD RIGHT HAND	.500
LEAD ANGLE	16°
ADDENDUM	.175
WHOLE DEPTH	.250
CHORDAL THICKNESS	.163

⑦ WORM GEAR
MATL: PHOSPHOR BRONZE

SPUR GEAR DATA	
DIAMETRAL PITCH	8
NUMBER OF TEETH	27
PRESSURE ANGLE	20°
PITCH DIAMETER	3.375
BASE CIRCLE DIAMETER	3.6187
CIRCULAR PITCH	.3927
CIRCULAR THICKNESS	.1964
ROOT DIAMETER	3.125

FIGURE 3-43 *continued*

PROBLEM 3-13 FLASHLIGHT

Materials

☐ AA battery

☐ Calipers

Directions for Problem 3-13

This problem requires you to develop a series of 3D solids models of the flashlight design. There are two purchase parts that are missing: the lightbulb and the battery.

Your task is to create the 10 parts. The AA battery will need to be created through reverse engineering using calipers and a sample battery. The lightbulb does not have specific dimensional details. See Figure 3-44. Use the chart provided to assist in the creation of a flashlight bulb. The actual bulb design, except for dimensions, will not be evaluated.

© 2010 Cengage Learning. All Rights Reserved. May not be scanned, copied or duplicated, or posted to a publicly accessible website, in whole or in part.

The remaining parts are detailed in the drawings provided.

Order of sheet development	PART	
1	BODY – BARREL	
2	TOP	
3	LENS	
4	CAP	
5	REFLECTOR	
6	METAL CLIP	
7	BULB HOLDER	
8	SPRING	
9	BATTERY	
10	BULB–Use Chart for guidance	
11	Exploded Assembly with Parts List	

Bulb information

TL-1-3/4	Midget Groove Base	MOD 0.19 inches (4.7 mm) MOL 0.69 inches (17.5 mm)
MOD = maximum overall diameter (glass) MOL = maximum overall length of the bulb (glass and base)		

© 2010 Cengage Learning. All Rights Reserved. May not be scanned, copied or duplicated, or posted to a publicly accessible website, in whole or in part.

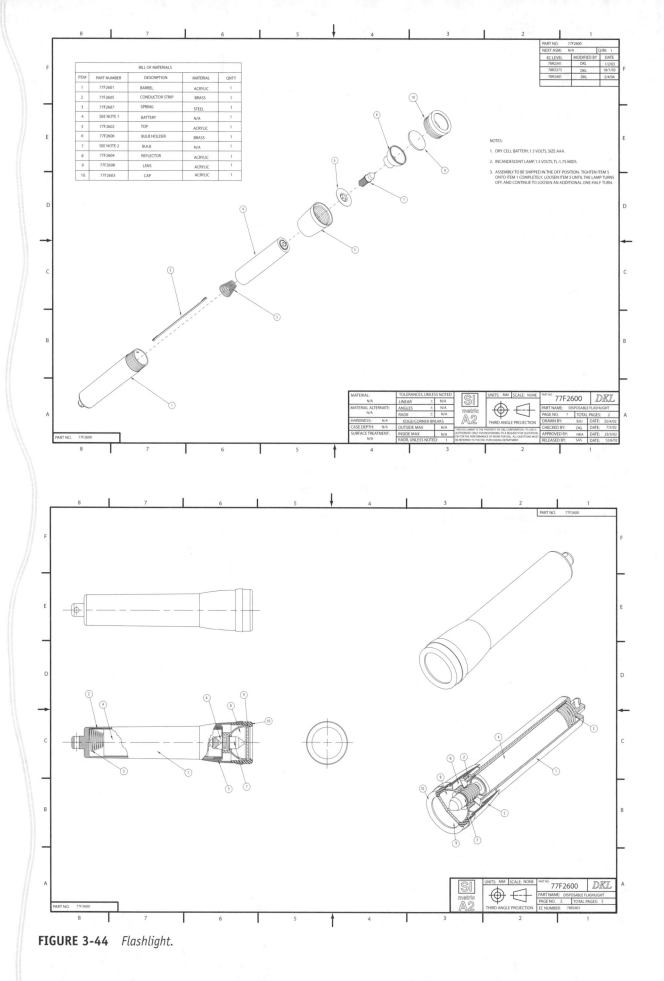

FIGURE 3-44 *Flashlight.*

© 2010 Cengage Learning. All Rights Reserved. May not be scanned, copied or duplicated, or posted to a publicly accessible website, in whole or in part.

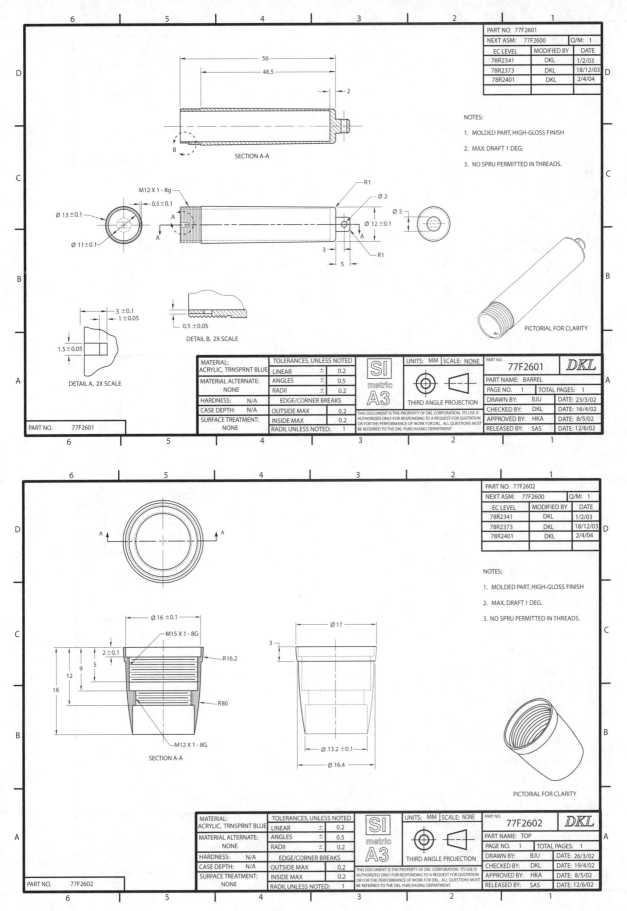

FIGURE 3-44 *continued*

© 2010 Cengage Learning. All Rights Reserved. May not be scanned, copied or duplicated, or posted to a publicly accessible website, in whole or in part.

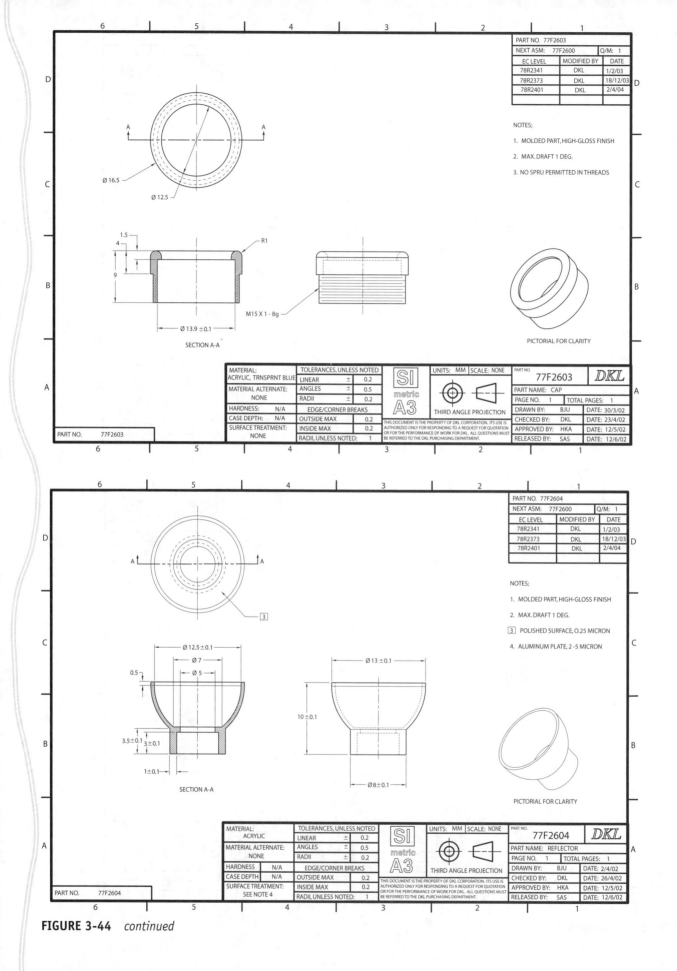

FIGURE 3-44 *continued*

© 2010 Cengage Learning. All Rights Reserved. May not be scanned, copied or duplicated, or posted to a publicly accessible website, in whole or in part.

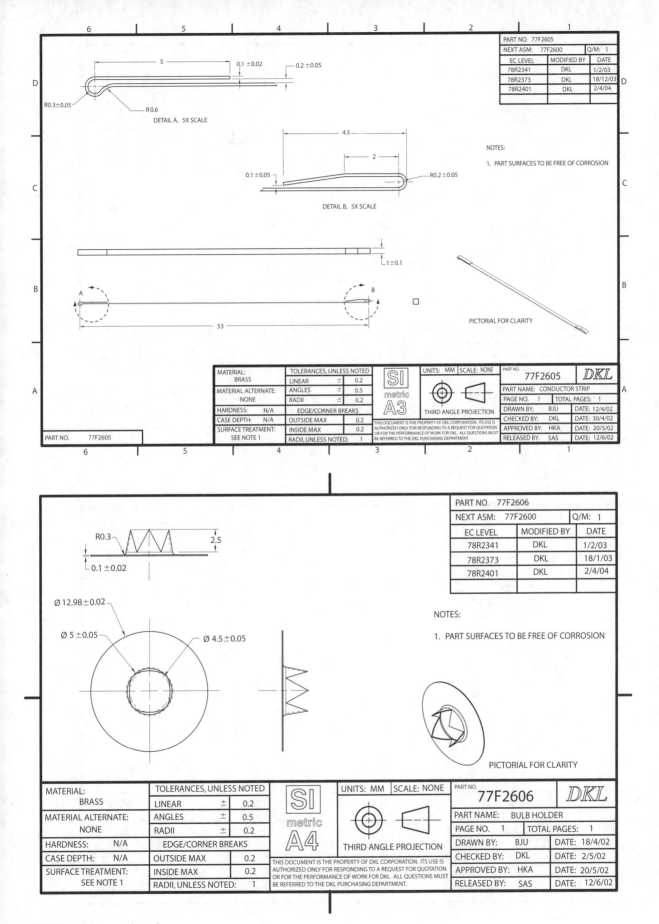

FIGURE 3-44 *continued*

© 2010 Cengage Learning. All Rights Reserved. May not be scanned, copied or duplicated, or posted to a publicly accessible website, in whole or in part.

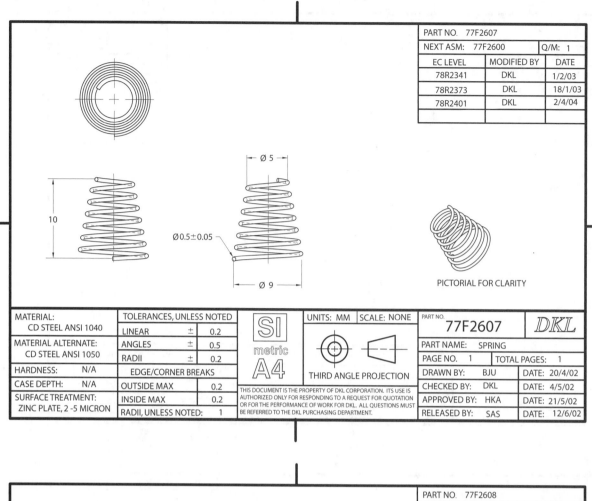

PART NO. 77F2607		
NEXT ASM: 77F2600		Q/M: 1
EC LEVEL	MODIFIED BY	DATE
78R2341	DKL	1/2/03
78R2373	DKL	18/1/03
78R2401	DKL	2/4/04

PICTORIAL FOR CLARITY

Ø 5
Ø0.5±0.05
Ø 9
10

MATERIAL: CD STEEL ANSI 1040	TOLERANCES, UNLESS NOTED			UNITS: MM	SCALE: NONE	PART NO. 77F2607	DKL
	LINEAR	±	0.2				
MATERIAL ALTERNATE: CD STEEL ANSI 1050	ANGLES	±	0.5			PART NAME: SPRING	
	RADII	±	0.2	THIRD ANGLE PROJECTION		PAGE NO. 1	TOTAL PAGES: 1
HARDNESS: N/A	EDGE/CORNER BREAKS					DRAWN BY: BJU	DATE: 20/4/02
CASE DEPTH: N/A	OUTSIDE MAX		0.2	THIS DOCUMENT IS THE PROPERTY OF DKL CORPORATION. ITS USE IS AUTHORIZED ONLY FOR RESPONDING TO A REQUEST FOR QUOTATION OR FOR THE PERFORMANCE OF WORK FOR DKL. ALL QUESTIONS MUST BE REFERRED TO THE DKL PURCHASING DEPARTMENT.		CHECKED BY: DKL	DATE: 4/5/02
SURFACE TREATMENT: ZINC PLATE, 2 -5 MICRON	INSIDE MAX		0.2			APPROVED BY: HKA	DATE: 21/5/02
	RADII, UNLESS NOTED:		1			RELEASED BY: SAS	DATE: 12/6/02

SI metric A4

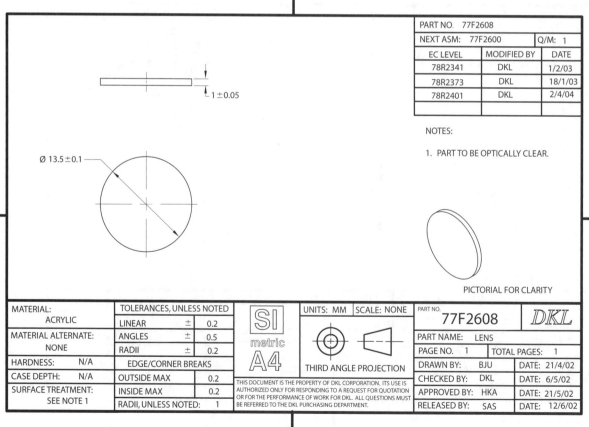

PART NO. 77F2608		
NEXT ASM: 77F2600		Q/M: 1
EC LEVEL	MODIFIED BY	DATE
78R2341	DKL	1/2/03
78R2373	DKL	18/1/03
78R2401	DKL	2/4/04

NOTES:

1. PART TO BE OPTICALLY CLEAR.

1±0.05
Ø 13.5±0.1

PICTORIAL FOR CLARITY

MATERIAL: ACRYLIC	TOLERANCES, UNLESS NOTED			UNITS: MM	SCALE: NONE	PART NO. 77F2608	DKL
	LINEAR	±	0.2				
MATERIAL ALTERNATE: NONE	ANGLES	±	0.5			PART NAME: LENS	
	RADII	±	0.2	THIRD ANGLE PROJECTION		PAGE NO. 1	TOTAL PAGES: 1
HARDNESS: N/A	EDGE/CORNER BREAKS					DRAWN BY: BJU	DATE: 21/4/02
CASE DEPTH: N/A	OUTSIDE MAX		0.2	THIS DOCUMENT IS THE PROPERTY OF DKL CORPORATION. ITS USE IS AUTHORIZED ONLY FOR RESPONDING TO A REQUEST FOR QUOTATION OR FOR THE PERFORMANCE OF WORK FOR DKL. ALL QUESTIONS MUST BE REFERRED TO THE DKL PURCHASING DEPARTMENT.		CHECKED BY: DKL	DATE: 6/5/02
SURFACE TREATMENT: SEE NOTE 1	INSIDE MAX		0.2			APPROVED BY: HKA	DATE: 21/5/02
	RADII, UNLESS NOTED:		1			RELEASED BY: SAS	DATE: 12/6/02

SI metric A4

FIGURE 3-44 *continued*

© 2010 Cengage Learning. All Rights Reserved. May not be scanned, copied or duplicated, or posted to a publicly accessible website, in whole or in part.

PROBLEM 3-14 CLAMP ASSEMBLY PROJECT

Directions for Problem 3-14

Create the press clamp assembly's individual parts and assemble it. Design a tool that clamps the material you need to hold in place. The clamping tool will vary based on the material you select. Provide written justification for the design choices of the clamp that were made on the clamping tool design.

Design deliverables: you need to clamp down a solid body electric guitar for a machining process; the body material is 2.25 inches tall. The second item is a golf club putter head (the one designed in the last chapter). Create a clamp to hold the object in place. You will need to design an end piece for each of the items to clamp down. Detail working drawings of the end piece component, assembly drawings, and a presentation-ready PowerPoint used to sell your design to potential manufacturers/funding sources. See Figure 3-45.

Advanced Level: Using drive constraints, make the clamp operational.

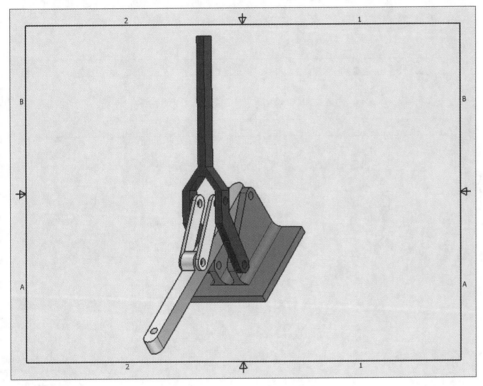

FIGURE 3-45(A) *Clamp.*

© 2010 Cengage Learning. All Rights Reserved. May not be scanned, copied or duplicated, or posted to a publicly accessible website, in whole or in part.

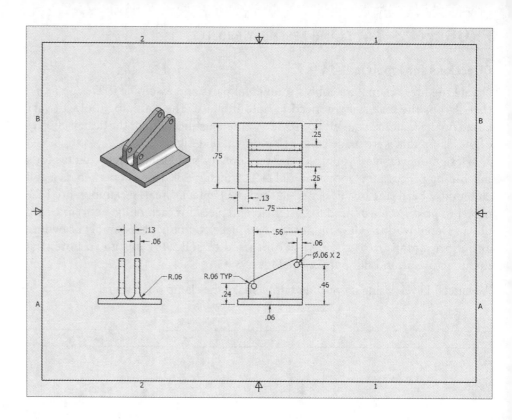

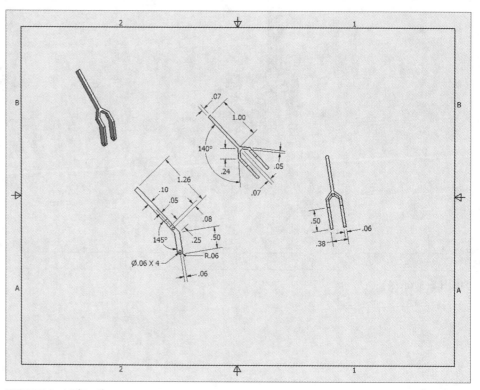

FIGURE 3-45(B–C)

© 2010 Cengage Learning. All Rights Reserved. May not be scanned, copied or duplicated, or posted to a publicly accessible website, in whole or in part.

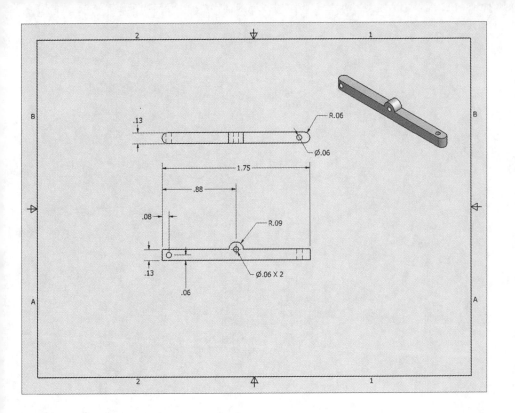

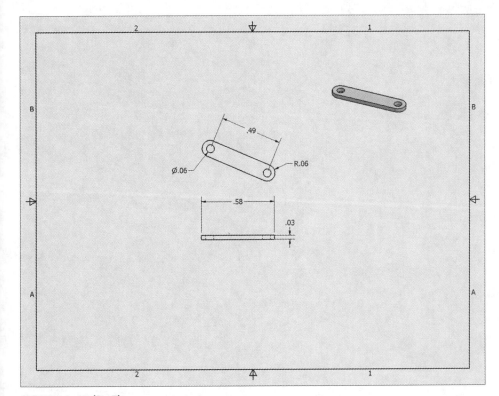

FIGURE 3-45(D–E)

© 2010 Cengage Learning. All Rights Reserved. May not be scanned, copied or duplicated, or posted to a publicly accessible website, in whole or in part.

UNIT 4
Advanced Manufacturing

Skills List

After completing the activities in this unit, you should be able to:

- Use a design for manufacturability verification sheet on a manufacturing process

- Develop an STL file for modeling

- Understand the different rapid prototyping processes

- Obtain a price quote for a prototype part

- Communicate using virtual teaming and VoIP technology

- Solve an open-ended design problem

© 2010 Cengage Learning. All Rights Reserved. May not be scanned, copied or duplicated, or posted to a publicly accessible website, in whole or in part.

SECTION 1
Design for Manufacturability

Design for Manufacturability

Use the following checklist of ideas to assist in narrowing design ideas into a final solution by applying the concepts of design for manufacturing.

Design Considerations

	Reduce the amount of packaging of the product
	Use stable structural shapes (triangle, circle) for load-bearing parts
	Try to reduce the bulk or weight of the product
	Use protective materials that are easy to apply and use
	Use reinforcement to strengthen high-stress locations
	If a part doesn't need to be separate, then combine it with another
	Look at ways to reduce the weight of the packaging
	Don't sacrifice quality for efficient design; look for a compromise
	Design with the future use in mind to increase the overall product life cycle
	Develop multifunctional use parts or assemblies
	Look for ways to combine construction (instead of having a cover, design the protective covering into the structure of the part)
	Develop location pins or guides to help ease and speed up the assembly of the parts

Environmentally Friendly Materials

	Use renewable energy resources in the production process
	Design for recyclable materials wherever possible
	Look for alternative substances in lieu of using toxic material
	Select materials with high recycled content over virgin materials
	Apply green manufacturing techniques by recycling resources used in the production of the product

© 2010 Cengage Learning. All Rights Reserved. May not be scanned, copied or duplicated, or posted to a publicly accessible website, in whole or in part.

	Avoid paint and stain. Powder coating and water-based paints have less environmental impact
	Design for alternative material sources if using a scarce resource or a monopolistic supplier. *Only one supplier available for a specific resource
	Eliminate the use of toxic substances or regulated materials if possible

Looking at "Green" Manufacturing

	Look for ways to eliminate waste in the manufacturing process
	Reclaim waste heat for preheating or another application
	Replace failed parts with high-efficiency motors and pumps
	Look at LED-based lighting systems for energy conservation
	Look at the big picture in the plant. Wsaste heat from one area could be used for hot water or other energy needs. Try to isolate areas of energy inefficiency

Improving Materials Handling

	Eliminate excessive movement or waiting time between operations of parts
	In automation-based systems, the use of counterweight adds to reduction power requirements to move materials
	Retain the potential energy of materials moved
	Minimize the number of operations required to assemble and improve the material supply aspect of the assembly area (eliminate excessive movements to assemble or locate parts)
	Use the space above your head to aid in material movement or drying

EXERCISE 4.1   APPLYING THE TECHNIQUES

1. Explain what design for manufacturing means?

2. How does design for manufacturing impact the product life cycle?

3. Select an item in your book bag and use the check sheet on design for manufacturability. Select one item in each section and apply the concept to improving how the part is made.

4. In teams of two students, visit one of the following Web sites and choose a video segment to review:

   a. The Science Channel's *How It's Made* Web site: http://science.discovery.com/fansites/howitsmade/howitsmade.html

   b. Stanford University's Alliance for Innovative Manufacturing Web site: http://manufacturing.stanford.edu/ Click on "How Everyday Things Are Made."

© 2010 Cengage Learning. All Rights Reserved. May not be scanned, copied or duplicated, or posted to a publicly accessible website, in whole or in part.

Typically, there are many manufacturing processes used in the development of the product. After you have reviewed a video segment, complete the following assignments:

a. Make a list of each process used, as indicated by the video.

b. Review the list of processes and identify three to five processes in the video that you think fit in a section of design for manufacturability.

c. Write a paragraph on how each of the processes would be improved by changing the design through applying the ideas in design for manufacturing.

© 2010 Cengage Learning. All Rights Reserved. May not be scanned, copied or duplicated, or posted to a publicly accessible website, in whole or in part.

SECTION 2
Rapid Prototyping Preparation Checklist

BACKGROUND

Rapid Prototyping

Rapid prototyping a design is an effective way to share proof of concept with others not tied directly to the design process. This process does require both equipment and supplies that a high school may not have. Alternative options to have prototype parts created includes partnering with a local community college or university.

Model Design Criteria

1. The size of the part (large or small) plays a role in selecting the correct machine for prototyping. If the part is thin-walled or small, considerations need to be made on the level of accuracy and the ability to build thin sections. A redesign or scaling may be needed before production.

2. Make sure your file is saved in both the native part format and the STL format. The STL (stereo lithography) format is a basic data format that indicates the part's structure to the software that is used to run the prototype machine. Note: Some CNC-based prototype machines use DXF file formats.

3. Check part twice run once. Verify the part's size and layout before you begin the build or cutting. It is easy to get caught up in the moment and forget to check to make sure the size in the prototype machine's software is correct.

Machine Setup

1. Check and clean the machine (remove scrap material) before setting up a part to run.

2. Run multiple parts at a time to maximize the build volume and speed of the machine (building two parts separately will take more time than if they are built at the same time).

3. Verify that there is enough material available to run the part(s).

4. Make sure the machine is calibrated through a test program or part if the machine has been idle for a long period of time.

© 2010 Cengage Learning. All Rights Reserved. May not be scanned, copied or duplicated, or posted to a publicly accessible website, in whole or in part.

EXERCISE 4.2   RAPID PROTOTYPING PROCESS AND PARTS WORKSHEET

Complete the following questions and submit a rapid prototyping quote request.

1. Describe the following types of rapid prototype part processes:
   a. SLA
   b. FDM
   c. SLS
   d. Cast Urethanes
   e. Rapid Injection Molding (Poly-Cast)

2. Name one major prototype equipment manufacturer for each of the following processes:
   a. SLA
   b. FDM
   c. SLS
   d. Cast Urethanes
   e. Rapid Injection Molding (Poly-Cast)

3. Calculate the estimate of the part cost
   a. Calculate the build volume of the part using mass properties.
   b. Estimate the cost of the actual part (not including support materials) using an estimated cost of $5 per cubic inch for material.

4. Submit a single part that was recently designed by you to the following Web site (www.quickparts.com) to obtain an automated online quote for all of the previous processes. Create a free e-mail account specifically for the quote registration, not your personal one. Compare each of the processes' time and cost information.

   Overall Part Dimensions

   X

   Y

   Z

   Volume

   _____

   a. SLA
      i. Time
      ii. Cost
      iii. Material
   b. FDM
      i. Time
      ii. Cost
      iii. Material
   c. SLS
      i. Time
      ii. Cost
      iii. Material

© 2010 Cengage Learning. All Rights Reserved. May not be scanned, copied or duplicated, or posted to a publicly accessible website, in whole or in part.

    d. Urethane

       i. Time

      ii. Cost

      iii. Material

    e. Rapid Injection Mold

       i. Time

      ii. Cost

      iii. Material

5. In each of the processes, there were different materials included in the quote. Select two different materials from each of the following processes: SLA and SLS.

    a. Identify the material.

       i. SLA

         1.

         2.

      ii. SLS

         1.

         2.

    b. What are the general characteristic properties of the material (research online)?

       i. SLA

         1.

         2.

      ii. SLS

         1.

         2.

© 2010 Cengage Learning. All Rights Reserved. May not be scanned, copied or duplicated, or posted to a publicly accessible website, in whole or in part.

SECTION 3
Virtual Teaming

BACKGROUND

Virtual Teaming

Virtual teaming uses the Internet to communicate with other team members, with the goal to create a design between remote teams.

The process is simple: Mimic what is currently transpiring in the design community around the world. VoIP (voice over Internet protocol) is now being used as a key communication and meeting tool due to the low cost and high quality of the communication channel. Since the designers and the manufacturing team may reside in different cities, states, or even countries, communication is the key to getting a high-quality product manufactured on time.

The concept of the project is for teams, made up of multiple sites, to work together to solve and design a solution to an open-ended design problem. The teams communicate virtually through several different resources, using free software tools both in and out of the classroom. The result will be a solution(s) designed to solve the problem using the input and innovation from the multiple sites involved.

Tools for Virtual Teaming

A headset that covers both ears with a boom microphone works best for class and home situations. A webcam is optional but offers a more personalized experience.

Four levels of collaboration can be realized at little or no cost to the individual:

Level 1: E-mail. You can use free e-mail accounts like Gmail, MSN, or your school account. E-mail is the most common way for virtual teams to communicate.

Level 2: Text chatting. Tools such as AOL IM, Yahoo IM, Skype, MSN, and others enable communicate through a typed interface.

Level 3: Voice chatting. Skype, Yahoo!, AOL, MSN, and others have VoIP tools that enable free computer-to-computer calls and group voice chats. To add a higher level of personal connection, you can integrate with webcams to provide live video that is paired with the audio.

Level 4: Sharing desktop applications while voice chatting. The free tool Yugma (www.yugma.com) can be paired with Skype to provide a complete solution for sharing designs in a live setting.

TIP SHEET

All of these tools can be downloaded online, for free; however, there is always a concern with security of your computer system. For the tools shared, especially Skype and Yugma, do not install any additional software, and do not have any advertising as part of the usage interface. Always be careful of security risks and never reply to outside uninvited user contacts through any means of Internet communication.

© 2010 Cengage Learning. All Rights Reserved. May not be scanned, copied or duplicated, or posted to a publicly accessible website, in whole or in part.

Steps to virtual teaming

1. Use the design process! This is a must; teams that go on their own usually have major conflicts.

2. Initially plan out a timeline (and use it); the fastest way to failure is not having a plan to follow.

3. Learn about your team members before jumping into the project. Insights into likes and dislikes will help form the team.

4. Begin your design process.

   a. Set and hold a time on brainstorming and move onto the next step.

   b. Make sure the design is done a few days before the presentation is created so images can be captured for the final presentation.

5. Practice the presentation online. It is difficult not seeing the immediate reaction of people listening. So one key to success is to listen to tones and inflections in the voices when questions are asked.

6. A helpful tip is to rotate the team leader position so everyone has to take on a decision-making role.

TIP SHEET

Etiquette:

Many times this area is skipped over, but the use of technology communication tools etiquette is a key aspect of effective team design and management processes.

1. Remember there is another person on the other end of the e-mail, IM, or voice chat.

2. Team responsibilities are with all team members (there is no I in team).

3. The key to success in this type of exercise is listening, and then providing comments after the thoughts are presented.

4. Don't use slang, concatenated words, or hurtful language; so "U" make sure the thought is presented clearly.

5. BE ON TIME; others are depending on you. If you are late the team is late and the work may not get done. So if there are out-of-class online meetings, make sure you own up to the responsibility.

6. Hold one conversation at a time. Talking to someone online and then switching to someone locally becomes distracting. Be polite and finish one conversation before starting another.

7. Cultures and locations can greatly influence a design, and sometimes cause conflict. Respect the differences and always take an inquisitive approach to help improve the design.

© 2010 Cengage Learning. All Rights Reserved. May not be scanned, copied or duplicated, or posted to a publicly accessible website, in whole or in part.

EXERCISE 4.3   FLASH DRIVE NAME TAGS

Developed by Steve Brown and Tom Singer

Materials and Equipment

Per team of 2–4 people:

- Engineering documentation tools or sketching materials to document design changes that took place
- Measurement tools, such as calipers, and steel rulers to measure existing flash drives attachment points and designs
- Headset(s) used for VoIP communication software
- At least one flash drive to use as a design guide

Safety and Disposal

Participants should always wear safety glasses when engaging in laboratory activities. In addition, anyone using equipment should be familiar with the operating instructions provided by the manufacturer of that equipment and should follow safe and proper operating procedures at all times.

No special safety or disposal procedures are required.

Participant Prerequisites

This project is designed for participants who have developed the following competencies and/or have completed the following:

- Basic knowledge and use of a parametric design software
- Typing skills used for Internet chat communication
- Applying communication etiquette, voice and typed
- Having an imagination

Team member Pre-Assessment Sheet: Flash Drive Name Tag

Although this is not a test, your responses will help your instructor provide a learning experience that meets your needs and helps you attain your goals. Please write your responses in the space provided at the back of this book.

- Have you used parametric design tools before?
- What experience (if any) do you have using VoIP or instant messaging tools?
- Have you ever applied the design process to a project?
- What team leadership experience do you have?

Team Member Instruction Sheet: Flash Drive Name Tag

The biggest problem with flash drives is being able to identify the owner. There typically is no way to write your name on a flash drive. The solution is to have a flash drive name tag that provides an area to have a logo and write your name.

Listed below are the overall constraints of the project and the deliverables from each design team.

1. Design a solution to the problem of flash drive identification.
2. This product design must be completed as an interactive application of the design process.
3. The design must incorporate at least two separate components in an assembly.
4. Each collaborative team will create parts of the design solution.

© 2010 Cengage Learning. All Rights Reserved. May not be scanned, copied or duplicated, or posted to a publicly accessible website, in whole or in part.

5. The solution needs to have an area where the user can write/label/insert their personal information so the drive can be identified.

6. The solution must also have the ability to be customized with corporate or educational logos.

7. The solution should easily fit in a shirt/pants pocket or purse along with the flash drive.

8. The design should be adaptable to multiple types of flash drives to maximize sales potential.

The collaborative team will be making a PowerPoint presentation on their design process as if it were to be presented to a potential sales client. Each collaborative team will have one presentation.

Deliverables:

- Detailed drawings of the team's solution
- Prototype models of the final project solution
- Communications utilization between the team members to convey/share ideas, problems, and solutions
- Documentation that the design process was followed
- PowerPoint presentation documentation

Team Member Post-Assessment Sheet: Flash Drive Name Tag

Team Dynamics

The team accomplished the tasks required?

My role in the team was? Did this role change during the project?

Did the team develop and follow a plan? How would you change the plan now that you're done?

Was there effective team leadership?

Did both teams (local and remote) provide input into the final solution?

Communications

Did the communication tools work well?

What improvements could be made on the communication tools?

Overall

What was the single best thing you learned about the project?

What was the worst part of the project?

© 2010 Cengage Learning. All Rights Reserved. May not be scanned, copied or duplicated, or posted to a publicly accessible website, in whole or in part.

SECTION 4

Open-Ended Design Problems

EXERCISE 4.4   ELEVATOR PANEL DESIGN PROJECT

Designing a layout for an elevator panel for a 12-story office building may sound like an easy task. But think a little further. You have accessibility issues, button placement logistics, and space for door controls.

The button panel is made from brushed aluminum and is sized at 7 inches wide and 14 inches in height with a 1/8 inch thickness. You need to place the following items:

12 floor location buttons

1 close door button

1 open door button

1 stop button—Must be a unique style

1 call Button—Must be a unique style

1 speaker grill area of 3.5 inches × 1.5 inches

For accessibility purposes buttons need to be the largest possible. Unique-style buttons must be a different design from all other buttons (size, shape, height). Sketch out your thoughts first and get them approved (instructor, if requested) before beginning the CAD design process. Don't forget to review ADA consideration like Braille and button sizes. See Figure 4-1.

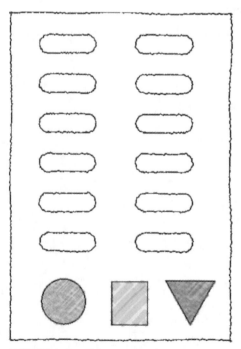

FIGURE 4-1 *Elevator panel.*

© 2010 Cengage Learning. All Rights Reserved. May not be scanned, copied or duplicated, or posted to a publicly accessible website, in whole or in part.

Deliverables

1. Sketch of design

2. Part models of the button, panel

3. Part drawings (select 1 floor button + the unique style buttons) and the panel with dimensions

4. Assembly of the panel with button placement

5. Exploded assembly with part's list

Making this a virtual teaming project:

1. This 12-story office tower is going to be built by a development company that has designated that the building will be built in Sarasota, FL (use as a doctor's office tower), Charlotte, NC (payment processing company), and Yuma, AZ (government services office building). They would like to have your recommendations on how the elevator panel could be designed for each application and if there are any design issues. Each team of designers must come up with application for the three building sites. Custom variations of the design are recommended.

2. Each team will research one of the cities, and perform a needs analysis for the site on which they are working. If there are two teams on the project the third site research will be split between the teams. Research should be done on the design in the following areas:

   a. Climate conditions in the city

   b. Building codes

   c. Average height of the residents (if it can be determined)

   d. Accessibility laws/rules

   e. Innovation in the elevator industry

3. A final presentation of the panel designs for each site will be done either locally or remotely using the virtual teaming tools.

EXERCISE 4.5   HANDS-FREE CELL PHONE HOLDER

Many states are enacting laws that require hands-free service for cell phone use while operating a moving vehicle. Bluetooth technology is great for linking your headset to your phone, but some phones don't have that. So, access to the phone may be necessary for dialing out. The problem is easy access to your phone. The goal will be to develop a cell phone holder that can be adapted for use in your car or office.

Research other devices with similar laws (GPS units) to get some ideas. Having a built-in solar–based charging system would be an additional plus.

It should be a universal holder that can accommodate a wide variety of cell phones.

Deliverables

1. Sketch of design

2. Part models

3. Part drawings

4. Assembly of the phone holder

5. Exploded assembly with part's list

© 2010 Cengage Learning. All Rights Reserved. May not be scanned, copied or duplicated, or posted to a publicly accessible website, in whole or in part.

Virtual Teaming Project

The hottest phone manufacturer (team to select phone mfg) on the market has contacted your design team to develop a cell phone holder. As a virtual group, select three of the phones from the manufacturer and develop an automobile cell phone holder that will work with three or more of the models. The goal is not to make it difficult to dial or start/end a call so think of ways to minimize the effect on driving.

Additional Deliverable for Virtual Teaming

Develop a presentation to sell your idea to potential investors/shareholders/ corporate buyers.

EXERCISE 4.6   KITCHEN DRAWER ORGANIZER FOR UTENSILS

Kitchen drawer organizers are used to separate and organize the utensils (forks, knives, spoons).

There is a variety of styles and materials used in the design.

Your task is to design a scalable utensil organizer, It must be able to fit drawers from 15 inches wide up to 30 inches wide (with additional add-ons). Drawer sizes vary from 15 inches, 18 inches, 21 inches, 24 inches, 27 inches, and 30 inches.

The internal structure must be customizable since the designs of the utensils will vary from manufacturer to manufacturer. Look for materials that will help resist the transfer of bacteria. The unit should be able to slide in and out of a drawer.

Deliverables

1. Sketch of design
2. Part models
3. Part drawings
4. Assembly of the drawer organizer
5. Exploded assembly with part's list
6. A presentation-ready PowerPoint used to sell your design to potential manufacturers/funding sources

EXERCISE 4.7   GOLF PUTTER DESIGN

A major golf club manufacturer has commissioned you to design the next unique golf putter. However, you have now realized that there are literally 1,000s of designs already done and many have been patented. Furthermore, there are specific rules that you need to follow, set forth by the USGA on putter design. Your task is to design a uniquely shaped or specially integrated shaft model of a putter. The putter could also be used to remove the ball from the cup after the hole is complete or have a unique alignment system to help make straight putts.

Deliverables

1. Sketch of design
2. Part models
3. Part drawings
4. Assembly of the drawer organizer
5. Exploded assembly with part's list

© 2010 Cengage Learning. All Rights Reserved. May not be scanned, copied or duplicated, or posted to a publicly accessible website, in whole or in part.

Virtual Teaming Project

Virtual teams could develop individual parts to the putter (head, shaft, grip) and assemble them for a completed assembly. Providing variation in the shaft design, grip options, and head design for a marketing presentation of potential investors into your company. (Having animations and rendered images is a must!)

Additional Deliverable for Virtual Teaming

Develop a presentation to sell you an idea to potential investors/shareholders/corporate buyers.

EXERCISE 4.8   MP3 SPEAKERS

Create the external design of a pair of MP3 speakers. The speaker size cannot exceed 4 inches (length $\times$ 5 inches height $\times$ 2 inches width). There must be a theme in the design of the speaker set which needs to be explained in the PowerPoint sales presentation to the class. Additional points are provided for "quick change" themes, which allow the user to change the theme of the speakers easily and quickly.

Deliverables

1. Sketch of design
2. Part models
3. Part drawings
4. Assembly of the drawer organizer
5. Exploded assembly with part's list

Virtual Teaming Project

Each team member will provide a sketch of a "quick change" mounting system to the rest of the team members so the team can decide on the best fastening system based on the criteria the team has determined.

Once the fastening system has been decided, each team member will design their own personal set of MP3 player speakers with a themed cover that used the attachment system.

The team members will then swap themed covers and in an assembly view show at least two additional variations of the themed covers besides their personal design.

Additional Deliverable for Virtual Teaming

A presentation of the team's best designed speakers and all the cover plates will be highlighted in a 10-minute presentation that will be one at the consumer electronics show as part of the product launch.

EXERCISE 4.9   "GREEN DESIGN" AUTOMATIC WATER LEVELING BIRD BATH

Design a bird bath that provides the ability to have a consistent water level as the water gets used or evaporates. The bird bath cannot exceed a 3 feet diameter size. There must be a supply of water that should last several days without human intervention. A pump can be used but it must be designed so that it can use solar photo-voltaic panels to power the pump. The collection of rainwater into a reservoir to make it a functional "green"-style system is a requirement.

© 2010 Cengage Learning. All Rights Reserved. May not be scanned, copied or duplicated, or posted to a publicly accessible website, in whole or in part.

Deliverables

1. Sketch of design
2. Part models
3. Part drawings
4. Assembly of the drawer organizer
5. Exploded assembly with part's list

Virtual Teaming Project

Each team would take a piece of the design. One on the reservoir system, one on the pumping system, and one on the bath system. The size of the bird bath will bow be 4 feet in diameter and be 2.75 inches deep at its lowest point. Some initial thoughts on communication items are: solar panel attachment system, pump placement and weatherproofing, bird bath mounting on the reservoir system

Additional Deliverable for Virtual Teaming

Develop a presentation to sell your idea to potential investors/shareholders/corporate buyers.

© 2010 Cengage Learning. All Rights Reserved. May not be scanned, copied or duplicated, or posted to a publicly accessible website, in whole or in part.

Appendix

Engineer's Notebook Pages and
Sketching Practice Sheets

© 2010 Cengage Learning. All Rights Reserved. May not be scanned, copied or duplicated, or posted to a publicly accessible website, in whole or in part.

241

Continued from page

SIGNATURE: DATE:

WITNESSED BY: DATE:

PROPRIETARY INFORMATION

241

© 2010 Cengage Learning. All Rights Reserved. May not be scanned, copied or duplicated, or posted to a publicly accessible website, in whole or in part.

Continued from page

Continued from page

SIGNATURE: DATE:

WITNESSED BY: DATE:

PROPRIETARY INFORMATION

242

© 2010 Cengage Learning. All Rights Reserved. May not be scanned, copied or duplicated, or posted to a publicly accessible website, in whole or in part.

Continued from page

243

Continued from page

SIGNATURE:

DATE:

WITNESSED BY:

DATE:

PROPRIETARY INFORMATION

© 2010 Cengage Learning. All Rights Reserved. May not be scanned, copied or duplicated, or posted to a publicly accessible website, in whole or in part.

Continued from page

Continued from page

SIGNATURE:

DATE:

WITNESSED BY:

DATE:

PROPRIETARY INFORMATION

244

© 2010 Cengage Learning. All Rights Reserved. May not be scanned, copied or duplicated, or posted to a publicly accessible website, in whole or in part.

Continued from page

Continued from page

SIGNATURE:

DATE:

WITNESSED BY:

DATE:

PROPRIETARY INFORMATION

245

© 2010 Cengage Learning. All Rights Reserved. May not be scanned, copied or duplicated, or posted to a publicly accessible website, in whole or in part.

Continued from page

Continued from page

SIGNATURE:

DATE:

WITNESSED BY:

DATE:

PROPRIETARY INFORMATION

246

© 2010 Cengage Learning. All Rights Reserved. May not be scanned, copied or duplicated, or posted to a publicly accessible website, in whole or in part.

Continued from page

247

Continued from page

SIGNATURE: DATE:

WITNESSED BY: DATE:

PROPRIETARY INFORMATION

247

© 2010 Cengage Learning. All Rights Reserved. May not be scanned, copied or duplicated, or posted to a publicly accessible website, in whole or in part.

Continued from page

Continued from page

SIGNATURE:

DATE:

WITNESSED BY:

DATE:

PROPRIETARY INFORMATION

248

© 2010 Cengage Learning. All Rights Reserved. May not be scanned, copied or duplicated, or posted to a publicly accessible website, in whole or in part.

Continued from page

PROPRIETARY INFORMATION

Continued from page

SIGNATURE:

DATE:

WITNESSED BY:

DATE:

249

© 2010 Cengage Learning. All Rights Reserved. May not be scanned, copied or duplicated, or posted to a publicly accessible website, in whole or in part.

Continued from page

Continued from page

SIGNATURE:

DATE:

WITNESSED BY:

DATE:

PROPRIETARY INFORMATION

250

© 2010 Cengage Learning. All Rights Reserved. May not be scanned, copied or duplicated, or posted to a publicly accessible website, in whole or in part.

Continued from page

251

Continued from page

SIGNATURE:

DATE:

WITNESSED BY:

DATE:

PROPRIETARY INFORMATION

© 2010 Cengage Learning. All Rights Reserved. May not be scanned, copied or duplicated, or posted to a publicly accessible website, in whole or in part.

Continued from page

Continued from page

SIGNATURE:

DATE:

WITNESSED BY:

DATE:

PROPRIETARY INFORMATION

252

© 2010 Cengage Learning. All Rights Reserved. May not be scanned, copied or duplicated, or posted to a publicly accessible website, in whole or in part.

Continued from page

253

Continued from page

SIGNATURE:

DATE:

WITNESSED BY:

DATE:

PROPRIETARY INFORMATION

© 2010 Cengage Learning. All Rights Reserved. May not be scanned, copied or duplicated, or posted to a publicly accessible website, in whole or in part.

Continued from page

Continued from page

SIGNATURE: DATE:

WITNESSED BY: DATE:

PROPRIETARY INFORMATION

254

© 2010 Cengage Learning. All Rights Reserved. May not be scanned, copied or duplicated, or posted to a publicly accessible website, in whole or in part.

Continued from page

Continued from page

SIGNATURE: DATE:

WITNESSED BY: DATE:

PROPRIETARY INFORMATION

255

© 2010 Cengage Learning. All Rights Reserved. May not be scanned, copied or duplicated, or posted to a publicly accessible website, in whole or in part.

Continued from page

Continued from page

SIGNATURE:

DATE:

WITNESSED BY:

DATE:

PROPRIETARY INFORMATION

256

© 2010 Cengage Learning. All Rights Reserved. May not be scanned, copied or duplicated, or posted to a publicly accessible website, in whole or in part.

Continued from page

257

Continued from page

SIGNATURE:

DATE:

WITNESSED BY:

DATE:

PROPRIETARY INFORMATION

257

© 2010 Cengage Learning. All Rights Reserved. May not be scanned, copied or duplicated, or posted to a publicly accessible website, in whole or in part.

Continued from page

Continued from page

SIGNATURE:

DATE:

WITNESSED BY:

DATE:

PROPRIETARY INFORMATION

258

© 2010 Cengage Learning. All Rights Reserved. May not be scanned, copied or duplicated, or posted to a publicly accessible website, in whole or in part.

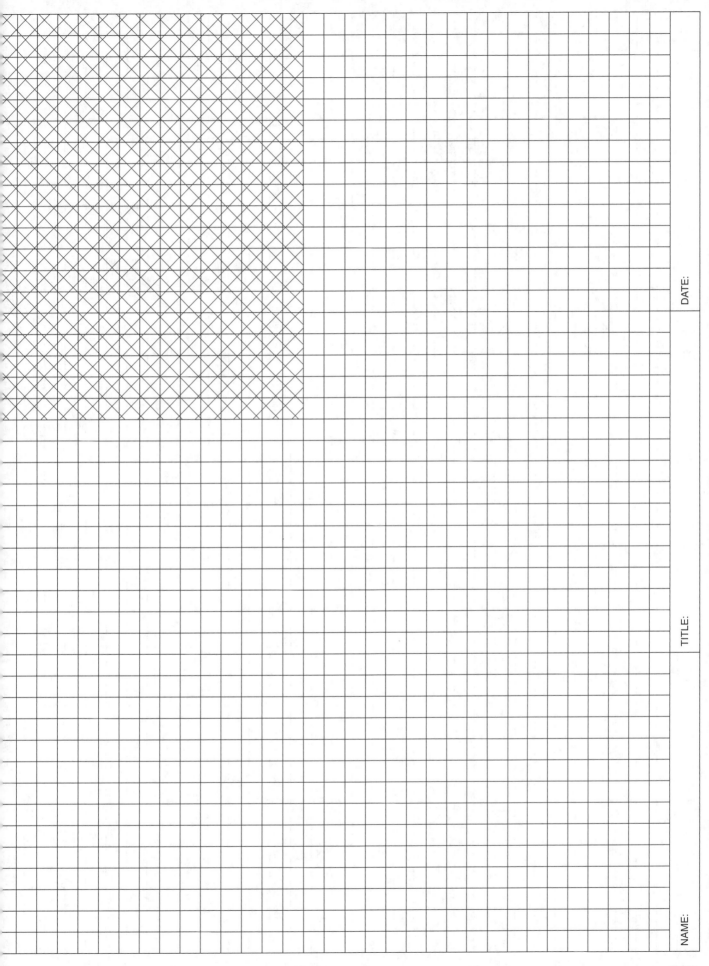

© 2010 Cengage Learning. All Rights Reserved. May not be scanned, copied or duplicated, or posted to a publicly accessible website, in whole or in part.

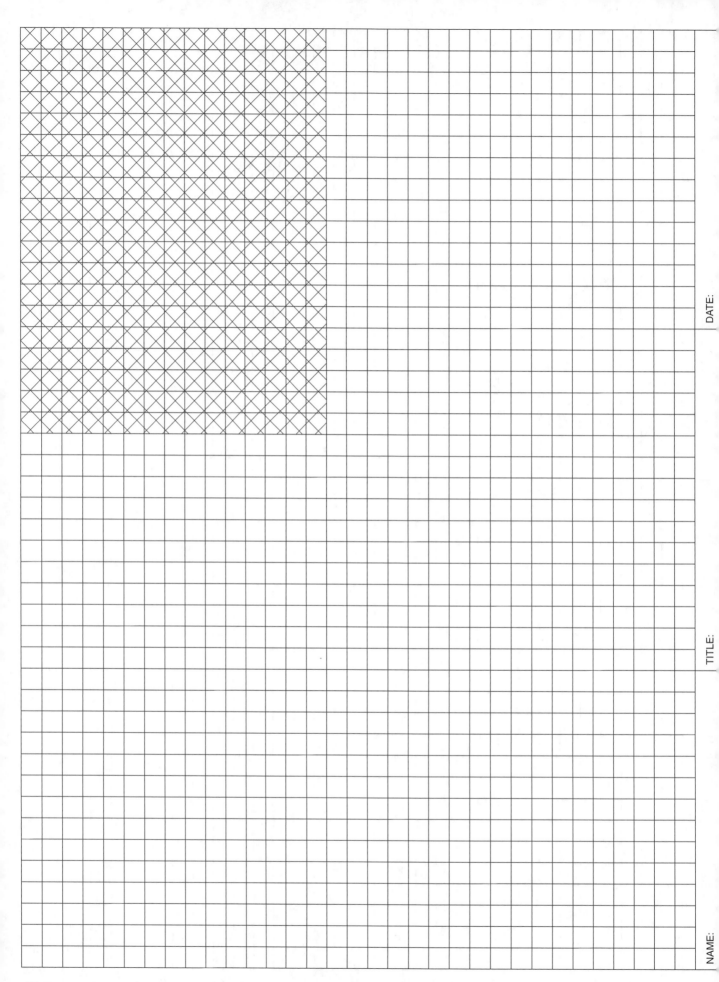

© 2010 Cengage Learning. All Rights Reserved. May not be scanned, copied or duplicated, or posted to a publicly accessible website, in whole or in part.

NAME:

TITLE:

DATE:

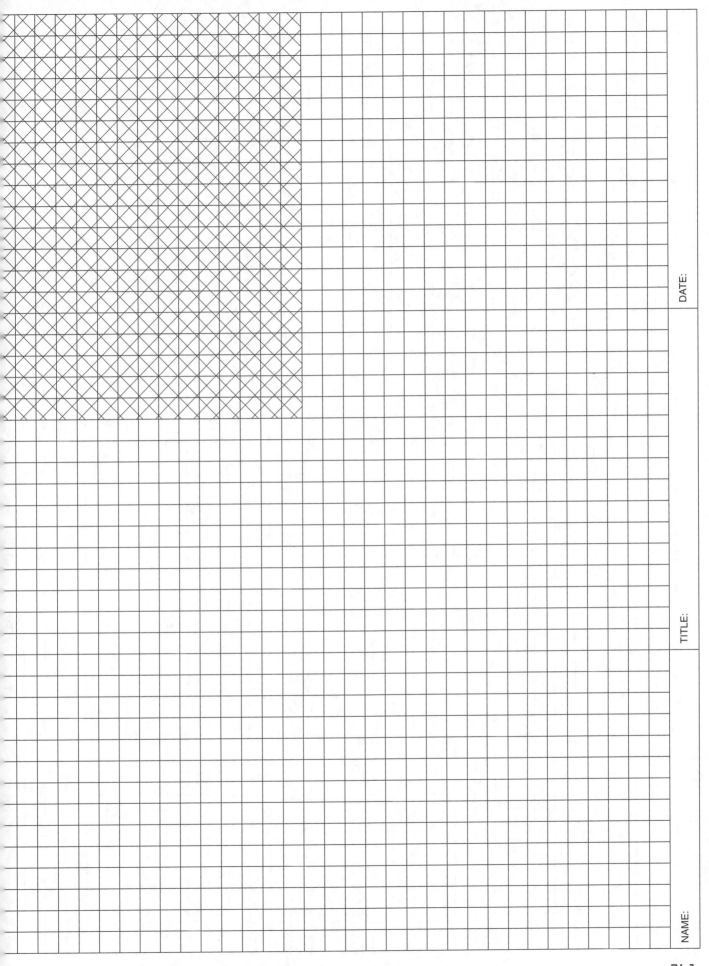

© 2010 Cengage Learning. All Rights Reserved. May not be scanned, copied or duplicated, or posted to a publicly accessible website, in whole or in part.

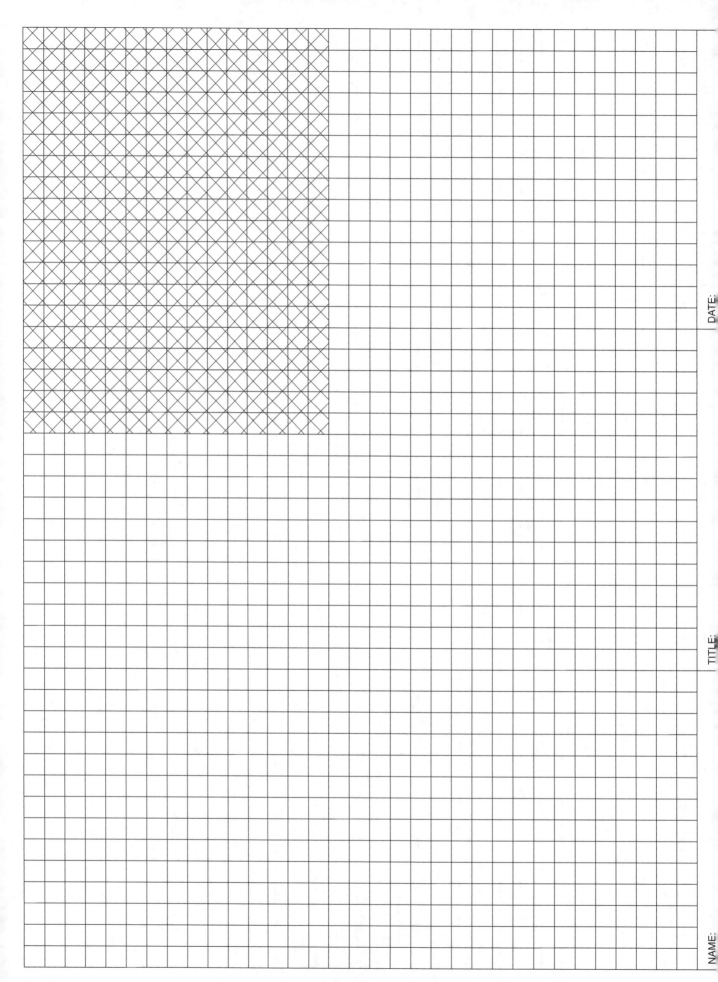

© 2010 Cengage Learning. All Rights Reserved. May not be scanned, copied or duplicated, or posted to a publicly accessible website, in whole or in part.

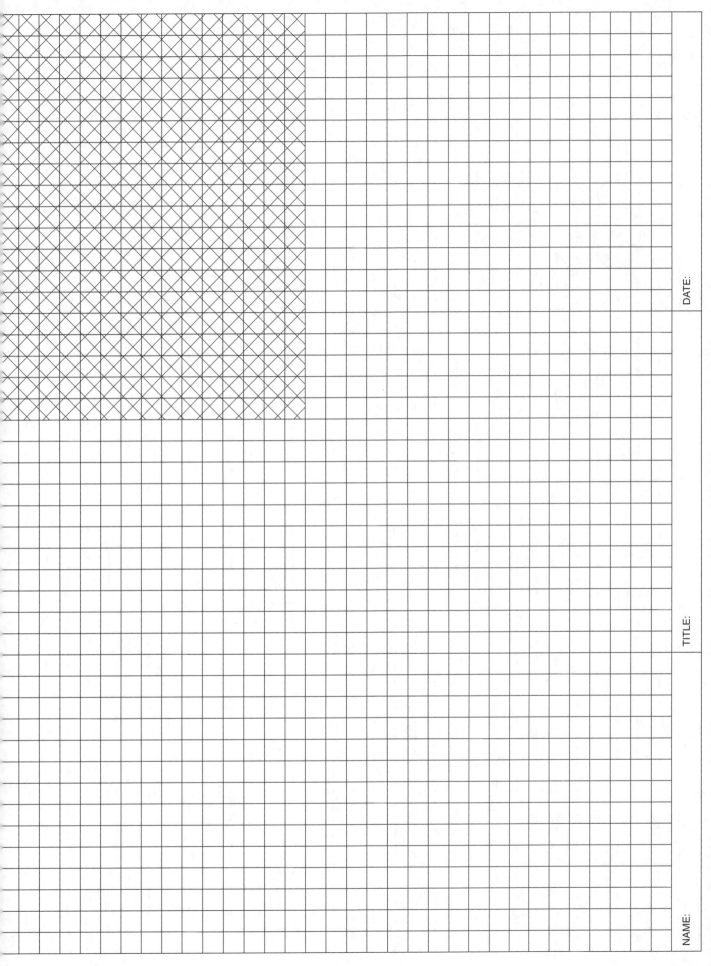

© 2010 Cengage Learning. All Rights Reserved. May not be scanned, copied or duplicated, or posted to a publicly accessible website, in whole or in part.

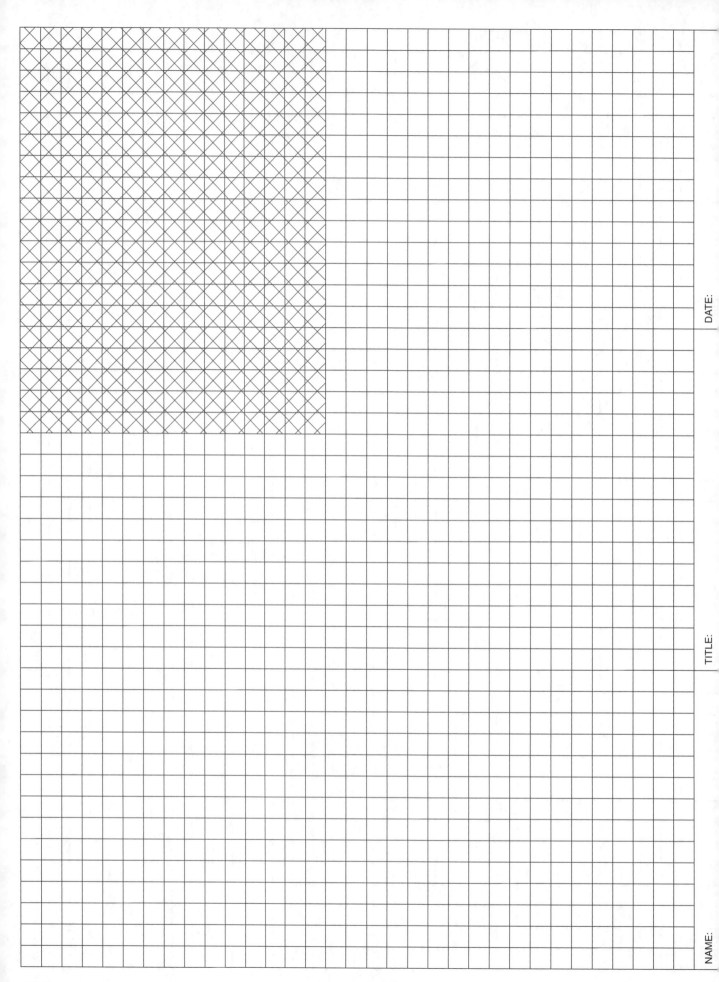

264

© 2010 Cengage Learning. All Rights Reserved. May not be scanned, copied or duplicated, or posted to a publicly accessible website, in whole or in part.

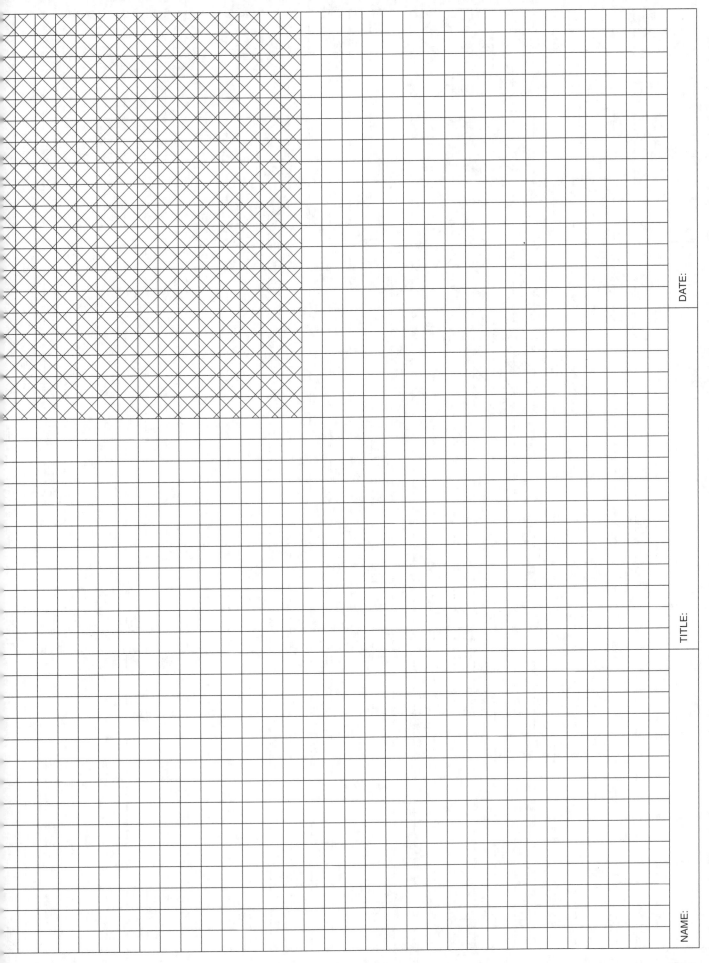

© 2010 Cengage Learning. All Rights Reserved. May not be scanned, copied or duplicated, or posted to a publicly accessible website, in whole or in part.

NAME:

TITLE:

DATE:

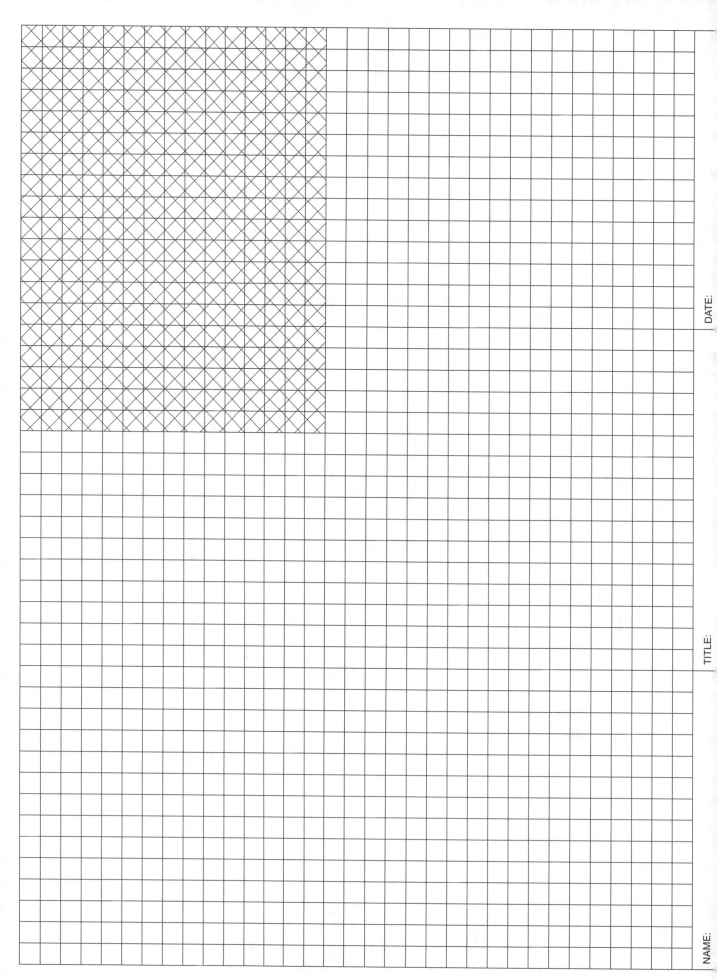

266

© 2010 Cengage Learning. All Rights Reserved. May not be scanned, copied or duplicated, or posted to a publicly accessible website, in whole or in part.

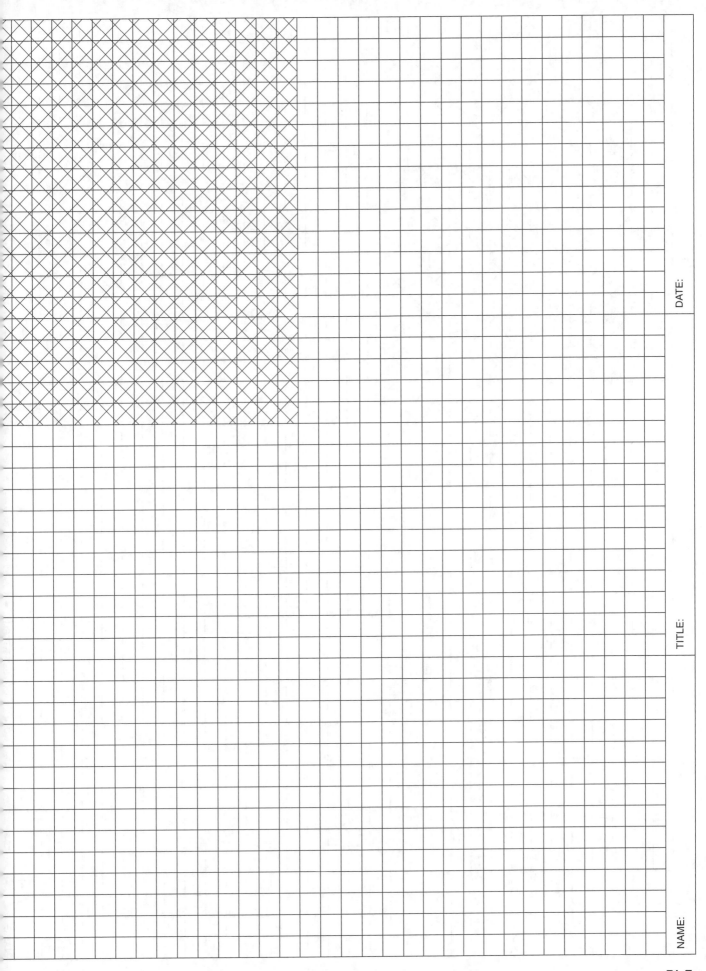

NAME:

TITLE:

DATE:

© 2010 Cengage Learning. All Rights Reserved. May not be scanned, copied or duplicated, or posted to a publicly accessible website, in whole or in part.

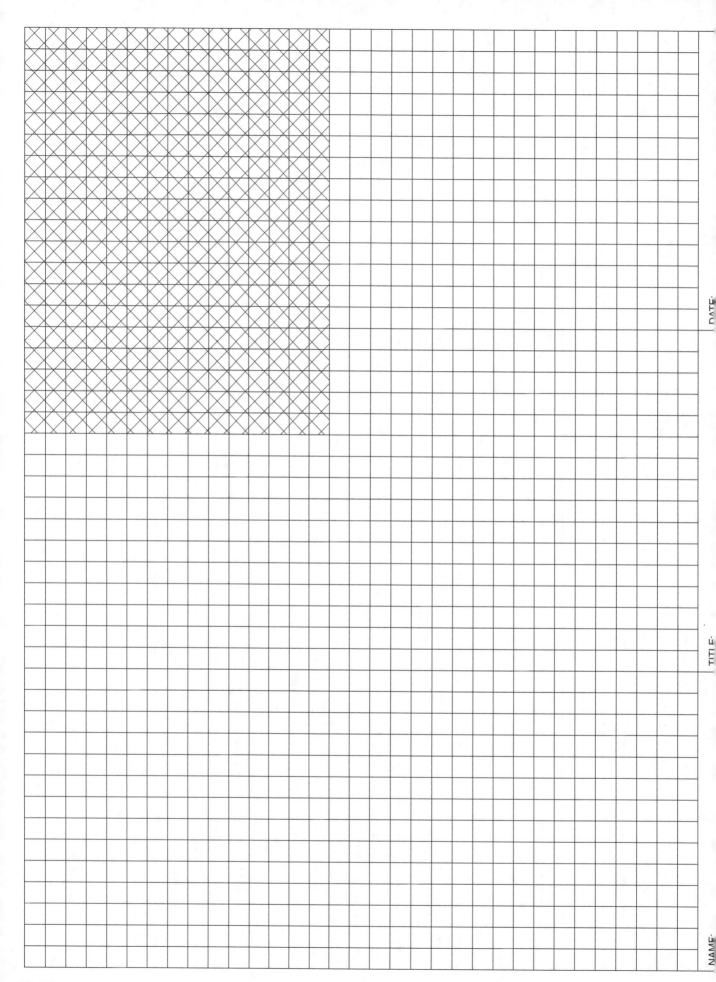

268

© 2010 Cengage Learning. All Rights Reserved. May not be scanned, copied or duplicated, or posted to a publicly accessible website, in whole or in part.

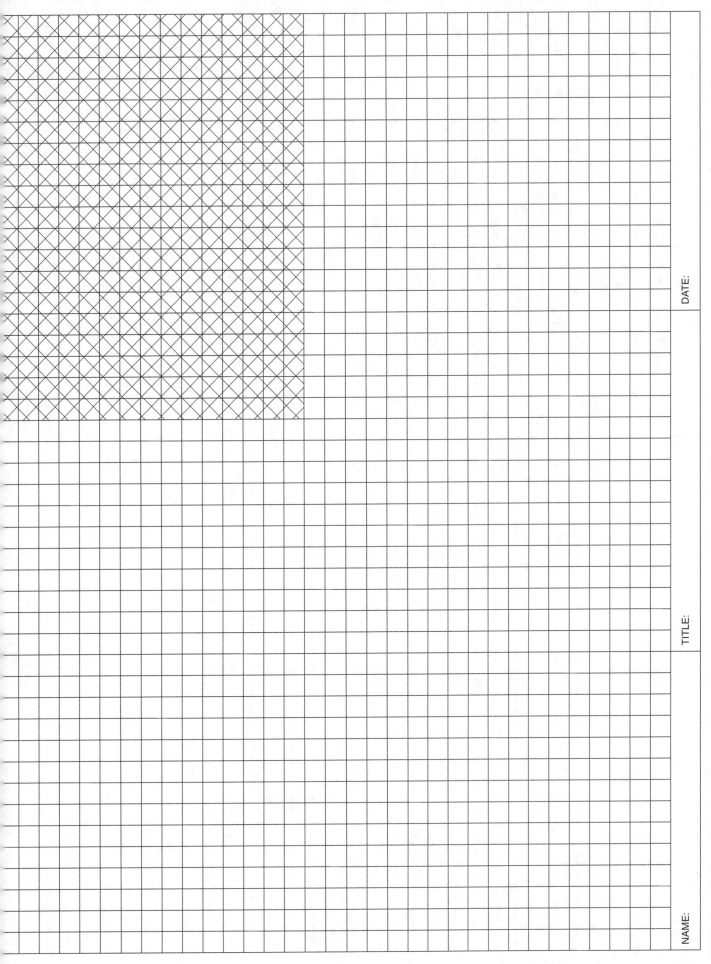

© 2010 Cengage Learning. All Rights Reserved. May not be scanned, copied or duplicated, or posted to a publicly accessible website, in whole or in part.

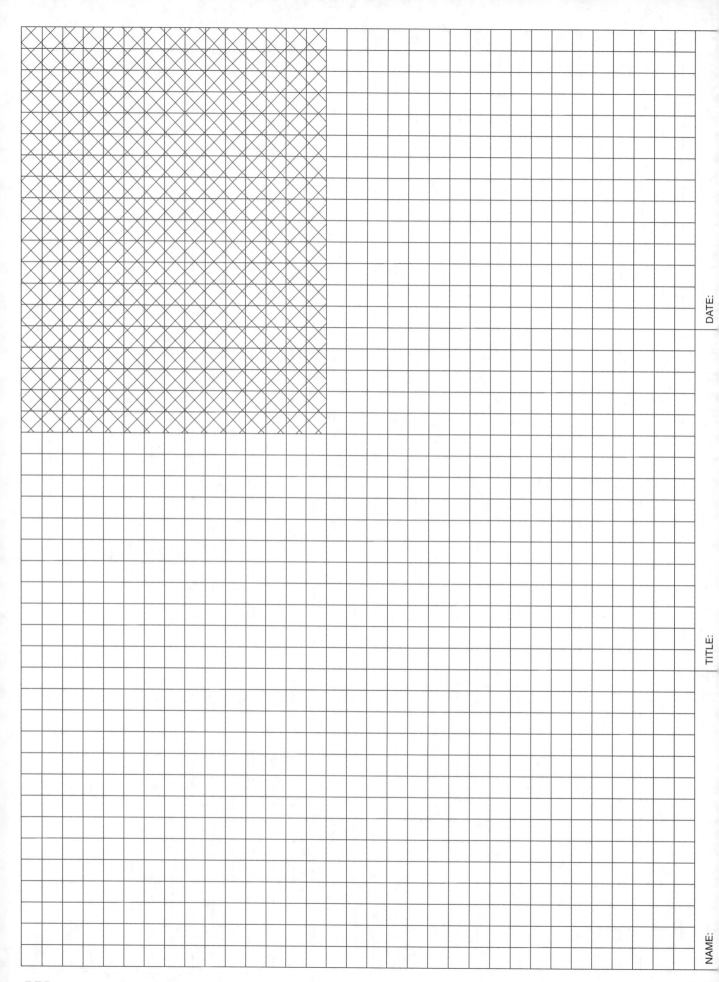

© 2010 Cengage Learning. All Rights Reserved. May not be scanned, copied or duplicated, or posted to a publicly accessible website, in whole or in part.

NAME:

TITLE:

DATE:

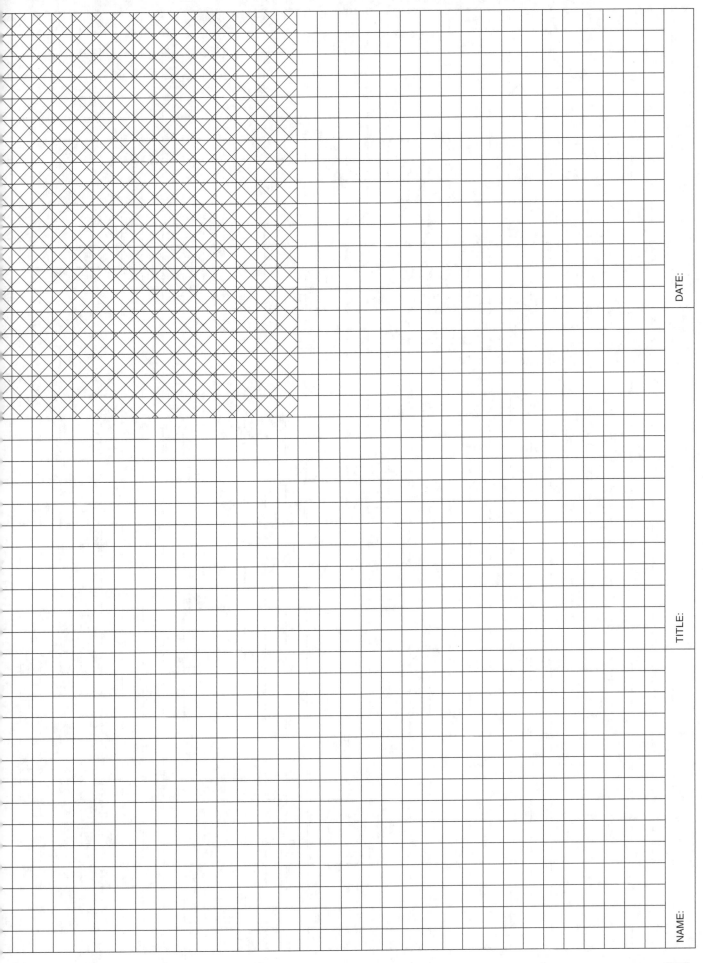

© 2010 Cengage Learning. All Rights Reserved. May not be scanned, copied or duplicated, or posted to a publicly accessible website, in whole or in part.

NAME: TITLE: DATE:

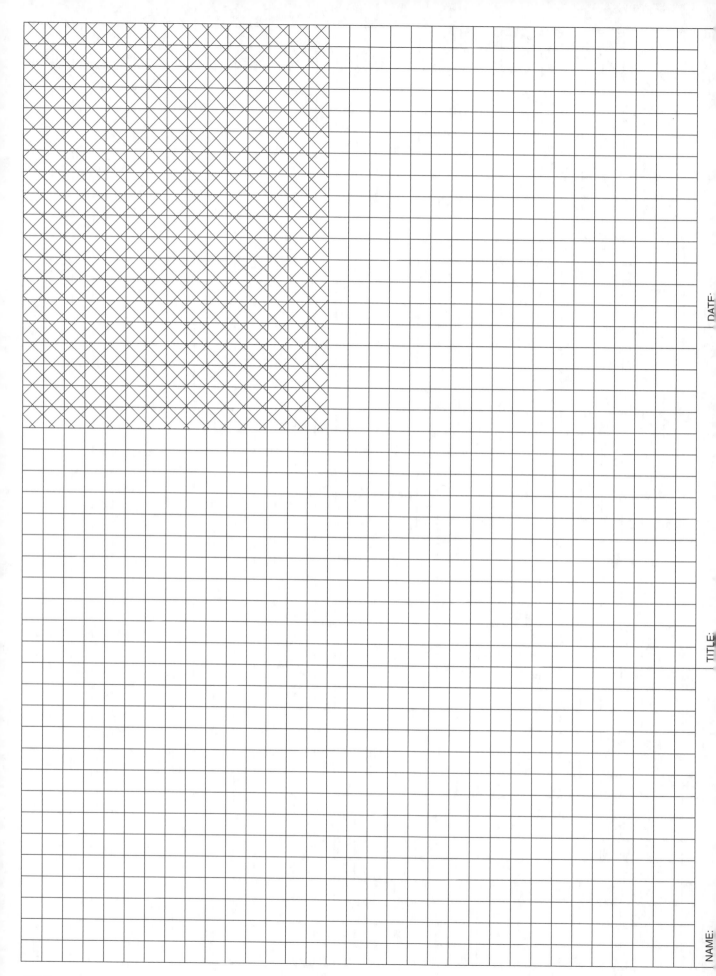

272

© 2010 Cengage Learning. All Rights Reserved. May not be scanned, copied or duplicated, or posted to a publicly accessible website, in whole or in part.

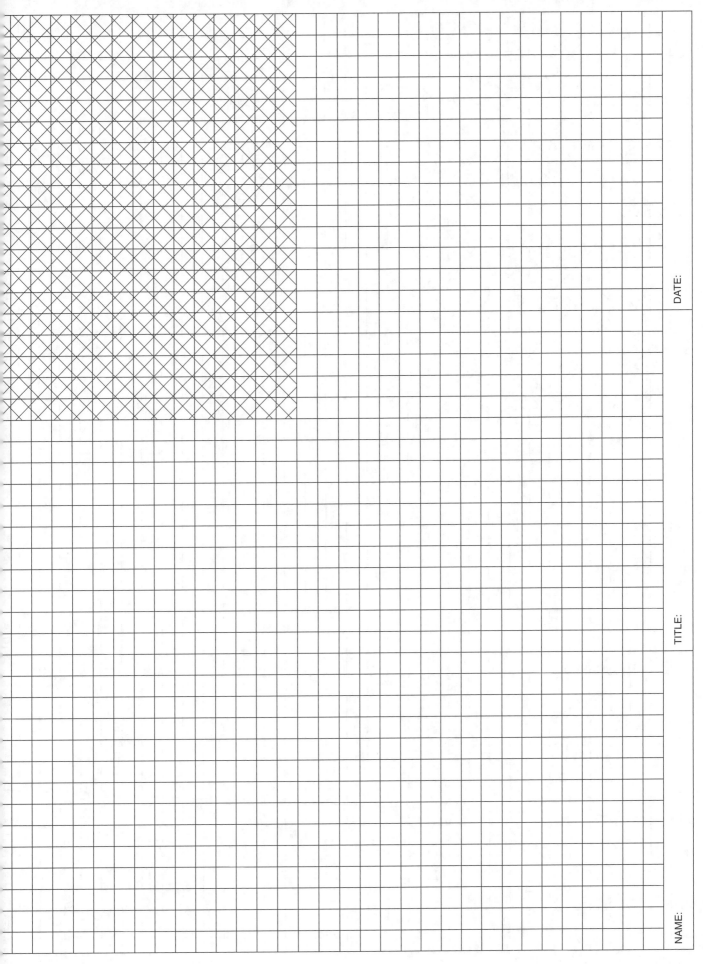

© 2010 Cengage Learning. All Rights Reserved. May not be scanned, copied or duplicated, or posted to a publicly accessible website, in whole or in part.

273

NAME:

TITLE:

DATE:

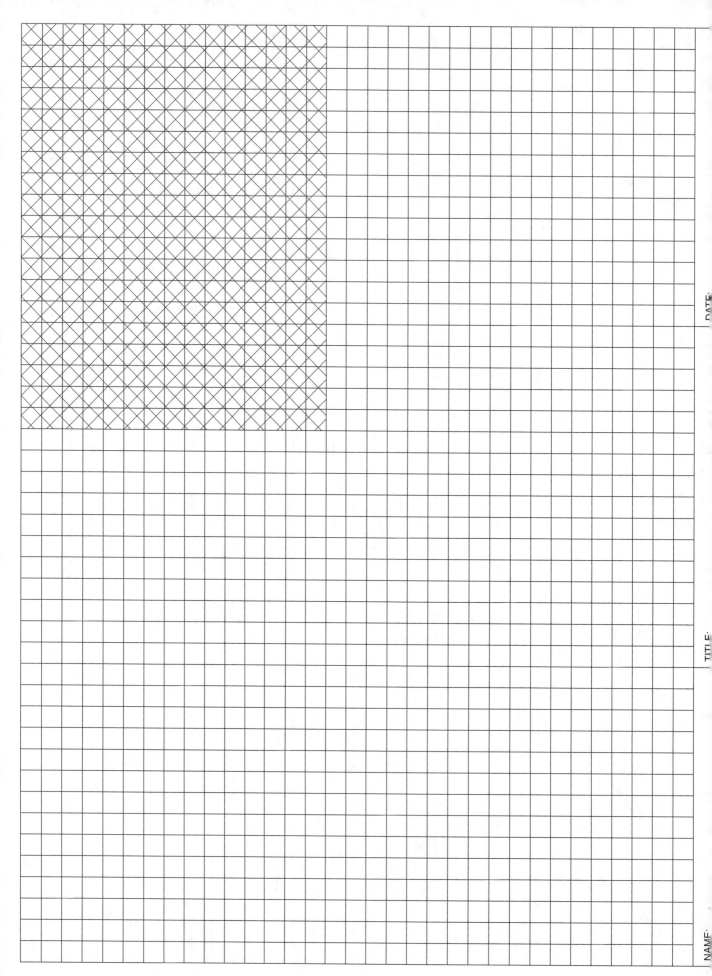

© 2010 Cengage Learning. All Rights Reserved. May not be scanned, copied or duplicated, or posted to a publicly accessible website, in whole or in part.

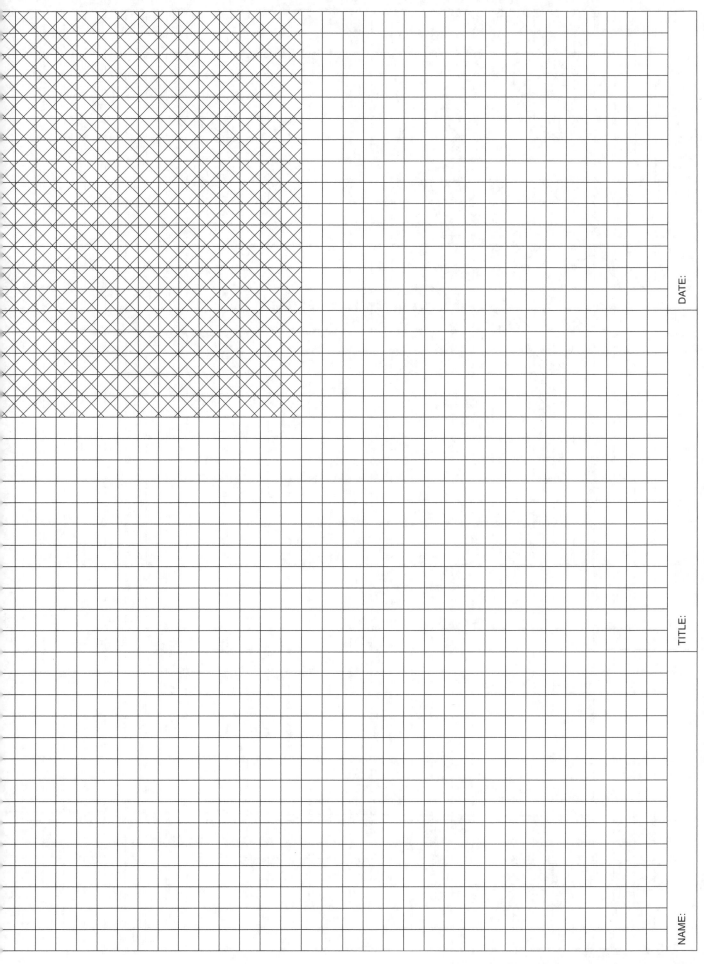

NAME:

TITLE:

DATE:

© 2010 Cengage Learning. All Rights Reserved. May not be scanned, copied or duplicated, or posted to a publicly accessible website, in whole or in part.

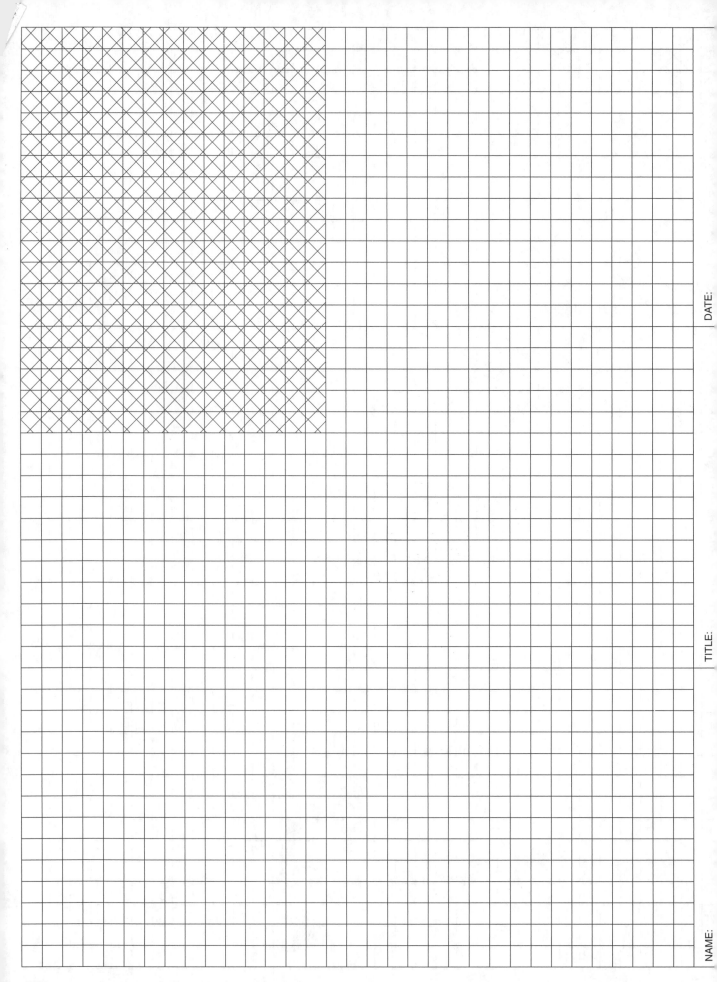

276

© 2010 Cengage Learning. All Rights Reserved. May not be scanned, copied or duplicated, or posted to a publicly accessible website, in whole or in part.

DATE:

TITLE:

NAME: